CURRENT DEVELOPMENTS IN MATHEMATICS, 1997

Editors:

Raoul Bott
Arthur Jaffe
David Jerison
George Lusztig
Isadore Singer
S. T. Yau

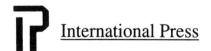 International Press

International Press Incorporated, Boston
P.O. Box 38-2872
Cambridge, MA 02238-2872

Post-conference Edition
ISBN 1-57146-078-0

Typeset using LaTeX
Printed on acid free paper, in the United States of America

Contents

1

PART I: LECTURES

TRACE FORMULA ON THE ADELE CLASS SPACE
AND WEIL POSITIVITY

Alain Connes

Department of Mathematics

Coll. de France

3 rue dUUlm

75005 Paris Cedex 05, FRANCE

connes@math.jussieu.fr

Abstract

We shall first show that the classification of factors, as seen in the unusual light of André Weil's Basic Number Theory, is a natural substitute for the Brauer theory of central simple algebras in local class field theory at Archimedian places. Passing to the global case provides a natural geometric framework in which the Frobenius, its eigenvalues and the Lefschetz formula interpretation of the explicit formulas continue to hold even for number fields. The geometric space involved is the Adele class space, i.e. the quotient of Adeles by the multiplicative group of the global field. We shall then explain that this leads to a natural spectral interpretation of the zeros of the Riemann zeta function and then prove the positivity of the Weil distribution assuming the validity of the analogue of the Selberg trace formula. The latter remains unproved and is equivalent to RH for all L-functions with Grössencharakter.

Introduction

Global fields k provide a natural context for the Riemann Hypothesis on the zeros of the zeta function and its generalization to Hecke L-functions. When the characteristic of k is non zero this conjecture was proved by A. Weil. His proof relies on the following dictionary (put in modern language) which gives a geometric meaning in terms of algebraic geometry over finite fields, to the function theoretic properties of the zeta functions. Recall that k is a function field over a curve Σ defined over $\mathbb{F}_q$,

Algebraic Geometry	*Function Theory*
Eigenvalues of action of Frobenius on $H_{et}^*(\overline{\Sigma}, \mathbb{Q}_\ell)$	Zeros and poles of ζ
Poincaré duality in ℓ-adic cohomology	Functional equation
Lefschetz formula for the Frobenius	Explicit formulas
Castelnuovo positivity	Riemann Hypothesis

We shall describe a third column in this dictionary, which will make sense for any global field. It is based on the geometry of the Adele class space,

(1) $$X = A/k^* , \ A = \text{Adeles of } k.$$

This space is of the same nature as the space of leaves of the horocycle foliation of a Riemann surface (section I) and the same geometry will be used to analyse it.

Our spectral interpretation of the zeros of zeta involves Hilbert space. The reasons why Hilbert space (apparently invented by Hilbert for this purpose) should be involved are manifold, let us mention three,

(A) The discovery of Hugh Montgomery ([M]) about the statistical fluctuations of the spacings of zeros of zeta. Numerical tests by Odlyzko ([O]) and further theoretical work by Katz-Sarnak ([KS]) give overwhelming evidence that zeros of zeta should be the eigenvalues of a hermitian matrix.

(B) The equivalence between RH and the positivity of the Weil distribution on the Idele class group C_k which shows that Hilbert space is implicitly present.

(C) The deep arithmetic significance of the work of A. Selberg on the spectral analysis of the Laplacian on $L^2(G/\Gamma)$ where Γ is an arithmetic subgroup of a semi simple Lie group G.

Direct atempts (cf. [B]) to construct the Polya-Hilbert space giving a spectral realization of the zeros of ζ using quantum mechanics, meet a serious minus sign problem explained in [B].

The very same $-$ sign appears in the Riemann-Weil explicit formula,

$$(2) \qquad \sum_{L(\chi,\rho)=0} \widehat{h}(\chi,\rho) - \widehat{h}(0) - \widehat{h}(1) = -\sum_v \int_{k_v^*}' \frac{h(u^{-1})}{|1-u|}\, d^*u\,,$$

where h is a test function on the Idele class group C_k, $\widehat{h}$ is its Fourier transform,

$$(3) \qquad \widehat{h}(\chi,z) = \int_{C_k} h(u)\,\chi(u)\,|u|^z\, d^*u\,,$$

and the finite values $\int'$ are suitably normalized. If we use the above dictionary when $\mathrm{char}(k) \neq 0$, the geometric origin of this $-$ sign becomes clear, the formula (2) is the Lefschetz formula,

$$(4) \qquad \#\text{ of fixed points of } \varphi = \mathrm{Trace}\,\varphi/H^0 - \mathrm{Trace}\,\varphi/H^1 + \mathrm{Trace}\,\varphi/H^2$$

in which the space $H^1_{\mathrm{et}}(\overline{\Sigma}, \mathbb{Q}_\ell)$ which provides the spectral realization of the zeros appears with a $-$ sign. This indicates that the spectral realization of zeros of zeta should be of cohomological nature or to be more specific, that the Polya-Hilbert space providing this realization should appear as the last term of an exact sequence of Hilbert spaces,

$$(5) \qquad 0 \to \mathcal{H}_0 \overset{T}{\to} \mathcal{H}_1 \to \mathcal{H} \to 0\,.$$

Let $X = A/k^*$ be the Adele class space. Our basic idea is to take for $\mathcal{H}_0$ a suitable completion of the codimension 2 subspace of functions on X such that,

$$(6) \qquad f(0) = 0\,, \quad \int f\, dx = 0\,,$$

while $\mathcal{H}_1 = L^2(C_k)$ and T is the restriction map coming from the inclusion $C_k \to X$, multiplied by $|a|^{1/2}$,

$$(7) \qquad (Tf)(a) = |a|^{1/2}\, f(a)\,.$$

The action of the Idele class group C_k which gives the spectral realization is then the obvious one, for $\mathcal{H}_0$

$$(8) \qquad (U(g)f)(x) = f(g^{-1}\,x) \qquad \forall\, g \in C_k$$

using the action by multiplication of C_k on X, and similarly the regular representation V for $\mathcal{H}_1$.

This idea works but there are two subtle points; first since X is a delicate quotient space the function spaces for X are naturally obtained by starting with function spaces on A and moding out by the "gauge transformations"

$$(9) \qquad\qquad f \to f_q, \; f_q(x) = f(xq), \quad \forall q \in k^*.$$

Here the natural function space is the Bruhat-Schwarz space $\mathcal{S}(A)$ and by (6) the codimension 2 subspace,

$$(10) \qquad\qquad \mathcal{S}(A)_0 = \left\{ f \in \mathcal{S}(A); \; f(0) = 0, \; \int f \, dx = 0 \right\}.$$

The restriction map T is then given by,

$$(11) \qquad\qquad T(f)(a) = |a|^{1/2} \sum_{q \in k^*} f(aq) \qquad \forall a \in C_k.$$

The corresponding function $T(f)$ belongs to $\mathcal{S}(C_k)$ and all functions $f - f_q$ are in the kernel of T.

The second subtle point is that since C_k is abelian and non compact, its regular representation does not contain any finite dimensional subrepresentation so that the Polya-Hilbert space cannot be a subrepresentation (or unitary quotient) of V. There is an easy way out (which will be improved later) which is to replace $L^2(C_k)$ by $L_\delta^2(C_k)$ using the polynomial weight $(\log^2 |a|)^{\delta/2}$, i.e. the norm,

$$(12) \qquad\qquad \|\xi\|_\delta^2 = \int_{C_k} |\xi(a)|^2 \, (1 + \log^2 |a|)^{\delta/2} \, d^*a.$$

Let $\mathrm{char}(k) = 0$ so that $\mathrm{Mod}\, k = \mathbb{R}_+^*$ and $C_k = K \times \mathbb{R}_+^*$ where K is the compact group $C_{k,1} = \{a \in C_k; \; |a| = 1\}$.

Theorem. ([Co]) *Let* $\delta > 1$, $\mathcal{H}$ *be the cokernel of* T *in* $L_\delta^2(C_k)$ *and* W *the quotient representation of* C_k. *Let* χ *be a character of* K, $\widetilde{\chi} = \chi \times 1$ *the corresponding character of* C_k. *Let* $\mathcal{H}_\chi = \{\xi \in \mathcal{H}; \; W(g)\,\xi = \chi(g)\,\xi \quad \forall g \in K\}$ *and* $D_\chi = \lim_{\epsilon \to 0} \frac{1}{\epsilon} (W(e^\epsilon) - 1)$. *Then* D_χ *is an unbounded closed operator with discrete spectrum,* $\mathrm{Sp}\, D_\chi \subset i\,\mathbb{R}$ *is the set of imaginary parts of zeros of the L function with Grössencharakter* $\widetilde{\chi}$ *which have real part 1/2. Moreover the spectral multiplicity of* ρ *is the largest integer* $n < \frac{1+\delta}{2}$ *in* $\{1, \dots, \text{multiplicity as a zero of } L\}$.

A similar result holds for $\mathrm{char}(k) > 0$. This allows to compute the character of the representation W as,

$$(13) \qquad\qquad \mathrm{Trace}(W(h)) = \sum_{\substack{L\left(x, \frac{1}{2}+\rho\right)=0 \\ \rho \in i\,\mathbb{R}/N^\perp}} \widehat{h}(\chi, \rho)$$

where $N = \mathrm{Mod}(k)$, $W(h) = \int W(g)\, h(g)\, d^*g$, $h \in \mathcal{S}(C_k)$, $\widehat{h}$ is defined in (3) and the multiplicity is counted as in the theorem.

This result is only preliminary because of the unwanted parameter δ which artificially restricts the multiplicities. The analogue of the Hodge $*$ operation is given on $\mathcal{H}_0$ by the Fourier transform,

$$(14) \qquad (Ff)(x) = \int_A f(y)\,\alpha(xy)\,dy \qquad \forall f \in \mathcal{S}(A)_0$$

which, because we take the quotient by (9), is independent of the choice of additive character α of A such that $\alpha \neq 1$ and $\alpha(q) = 1 \quad \forall q \in k$. Note also that $F^2 = 1$ on the quotient. On $\mathcal{H}_1$ the Hodge $*$ is given by,

$$(15) \qquad (*\xi)(a) = \xi(a^{-1}) \qquad \forall a \in C_k \,.$$

The Poisson formula means exactly that T commutes with the $*$ operation. (cf. section V for signs and powers of i). This is just a reformulation of the work of Tate and Iwasawa on the proof of the functional equation, but we shall now see that if we follow the proof by Atiyah-Bott ([AB]) of the Lefschetz formula we do obtain a clear geometric meaning for the Weil distribution. One can of course as in ([G]) define inner products on function spaces on C_k using the Weil distribution, but as long as the latter is put by hands and does not appear naturally one has very little chance to understand why it should be positive. Now, let φ be a diffeomorphism of a smooth manifold Σ and assume that the graph of φ is transverse to the diagonal, one can then easily define and compute (cf. [AB]) the distribution theoretic trace of the permutation U of functions on Σ associated to φ,

$$(16) \qquad (U\xi)(x) = \xi(\varphi(x)) \qquad \forall x \in \Sigma \,.$$

One has "Trace" $(U) = \int k(x,x)\,dx$, where $k(x,y)\,dy$ is the Schwarz kernel associated to U, i.e. the distribution on $\Sigma \times \Sigma$ such that,

$$(17) \qquad (U\xi)(x) = \int k(x,y)\,\xi(y)\,dy \,.$$

Now near the diagonal and in local coordinates one has,

$$(18) \qquad k(x,y) = \delta(y - \varphi(x)) \,,$$

where δ is the Dirac distribution. One then obtains,

$$(19) \qquad \text{"Trace"}\,(U) = \sum_{\varphi(x)=x} \frac{1}{|1 - \varphi'(x)|} \,,$$

where φ' is the Jacobian of φ and $|\,|$ stands for the absolute value of the determinant.

With more work ([GS]) one obtains a similar formula for the distributional trace of the action of a flow,

$$(20) \qquad (U_t\xi)(x) = \xi(F_t(x)) \qquad \forall x \in \Sigma,\ t \in \mathbb{R} \,.$$

It is given, under suitable transversality hypothesis, by

$$(21) \qquad \text{"Trace"} \ (U(h)) = \sum_\gamma \int_{I_\gamma} \frac{h(u)}{|1 - (F_u)_*|} \, d^*u \, ,$$

where $U(h) = \int h(t) \, U(t) \, dt$, h is a test function on $\mathbb{R}$, the γ labels the periodic orbits of the flow, including the fixed points, I_γ is the corresponding isotropy subgroup, and $(F_u)_*$ is the tangent map to F_u on the transverse space to the orbits, and finally d^*u is the unique Haar measure on I_γ which is of covolume 1 in $(\mathbb{R}, dt)$.

Now it is truly remarkable that when one analyzes the periodic orbits of the action of C_k on X one finds that (21) becomes,

$$(22) \qquad \text{"Trace"} \ (U(h)) = \sum_v \int_{k_v^*} \frac{h(u^{-1})}{|1 - u|} \, d^*u \, .$$

Thus, the isotropy subgroups I_γ are parametrized by the places v of k and coincide with the natural cocompact inclusion $k_v^* \subset C_k$ which relates local to global in class field theory. The denominator $|1 - u|$ is for the module of the local field k_v and the u^{-1} in $h(u^{-1})$ comes from the discrepancy between notations (8) and (16). It turns out that if one normalizes the Haar measure d^*u of modulated groups as in Weil ([W3]) , by,

$$(23) \qquad \int_{1 \leq |u| \leq \Lambda} d^*u \sim \log \Lambda \qquad \text{for } \Lambda \to \infty \, ,$$

one gets the same covolume 1 condition as in (21).

The transversality condition imposes the condition $h(1) = 0$. The distributional trace for the action of C_k on C_k by translations vanishes under the condition $h(1) = 0$.

Thus equating the alternate sum of traces on $\mathcal{H}_0$, $\mathcal{H}_1$ with the trace on the cohomology should thus provide the geometric understanding of the Riemann-Weil explicit formula (2) and in fact of RH using (13) if it could be justified for some value of δ.

The trace of permutation matrices is positive and this explains the Hadamard positivity,

$$(24) \qquad \text{"Trace"} \ (U(h)) \geq 0 \qquad \forall \, h \, , \ h(1) = 0 \, , \ h(u) \geq 0 \quad \forall \, u \in C$$

(not to be confused with Weil postivity).

To eliminate the artificial parameter δ and give rigorous meaning, as a Hilbert space trace, to the distribution "trace", one proceeds as in the Selberg trace formula and introduces a cutoff. In physics terminology the divergence of the trace is both infrared and ultraviolet as is seen in the simplest case of the action of K^* on $L^2(K)$ for a local field K. In this local case one lets,

$$(25) \qquad R_\Lambda = \widehat{P}_\Lambda \, P_\Lambda \, , \ \Lambda \in \mathbb{R}_+ \, ,$$

where P_Λ is the orthogonal projection on the subspace,

(26) $$\{\xi \in L^2(K)\,;\; \xi(x) = 0 \qquad \forall\, x, |x| > \Lambda\}\,,$$

while $\widehat{P}_\Lambda = F\,P_\Lambda\,F^{-1}$, F the Fourier transform.

One proves ([Co]) in this local case the following analogue of the Selberg trace formula,

(27) $$\text{Trace}\,(R_\Lambda\,U(h)) = 2\,h(1)\log'(\Lambda) + \int^{'} \frac{h(u^{-1})}{|1-u|}\,d^*u + o(1)$$

where $h \in \mathcal{S}(K^*)$ has compact support, $2\log'(\Lambda) = \int_{\lambda \in K^*,\,|\lambda| \in [\Lambda^{-1},\Lambda]} d^*\lambda$, and the principal value $\int^{'}$ is uniquely determined by the pairing with the unique distribution on K which agrees with $\frac{du}{|1-u|}$ for $u \neq 1$ and whose Fourier transform vanishes at 1.

As it turns out this principal value agrees with that of Weil for the choice of Fourier transform F associated to the standard character of K.

Let k be a global field and let first S be a finite set of places of k containing all the infinite places. To S corresponds the following localized version of the action of C_k on X. One replaces C_k by

(28) $$C_S = \prod_{v \in S} k_v^*/O_S^*\,,$$

where $O_S^* \subset k^*$ is the group of S-units. One replaces X by

(29) $$X_S = \prod_{v \in S} k_v/O_S^*\,.$$

The Hilbert space $L^2(X_S)$, its Fourier transform F and the orthogonal projections P_Λ, $\widehat{P}_\Lambda = F\,P_\Lambda\,F^{-1}$ continue to make sense, with

(30) $$\text{Im}\,P_\Lambda = \{\xi \in L^2(X_S)\,;\; \xi(x) = 0 \quad \forall\, x,\; |x| > \Lambda\}\,.$$

As soon as S contains more than 3 elements, (e.g. $\{2, 3, \infty\}$ for $k = \mathbb{Q}$) the space X_S is an extremely delicate quotient space. It is thus quite remarkable that the *trace formula* holds,

Theorem. ([Co]) *For any $h \in \mathcal{S}_c(C_S)$ one has, with $R_\Lambda = \widehat{P}_\Lambda\,P_\Lambda$,*

$$\text{Trace}\,(R_\Lambda\,U(h)) = 2\log'(\Lambda)\,h(1) + \sum_{v \in S} \int_{k_v^*}^{'} \frac{h(u^{-1})}{|1-u|}\,d^*u + o(1)$$

where the notations are as above and the finite values $\int^{'}$ depend on the additive character of Πk_v defining the Fourier transform F. When $\text{Char}\,(k) = 0$ the projectors P_Λ, $\widehat{P}_\Lambda$ commute on L^2_χ for Λ large enough so that one can replace R_Λ by the orthogonal projection Q_Λ on $\text{Im}\,P_\Lambda \cap \text{Im}\,\widehat{P}_\Lambda$. The situation for $\text{Char}\,(k) = 0$ is more delicate since P_Λ and $\widehat{P}_\Lambda$ do not commute (for Λ large) even in the local Archimedian case. But fortunately these operators commute ([LPS]) with a specific second order differential operator, whose eigenfunctions, the Prolate Spheroidal Wave functions provide the right filtration Q_Λ. This allows to replace R_Λ by Q_Λ and to state the global trace formula

$$\text{Trace}\,(Q_\Lambda\, U(h)) = 2\log'(\Lambda)\, h(1) + \sum_v \int_{k_v^*} \frac{h(u^{-1})}{|1-u|}\, d^*u + o(1)\,.$$

Our final result is that the validity of this trace formula implies (in fact is equivalent to) the positivity of the Weil distribution, i.e. *RH* for all *L*-functions with Grössencharakter. Moreover the filtration by Q_Λ allows to define the Adelic cohomology and to complete the dictionary between the function theory and the geometry of the Adele class space X.

Function Theory	*Geometry*
Zeros and poles of Zeta	Action of C_k on Adelic cohomology
Functional Equation	* operation
Explicit formulas	Lefschetz formula for action of C_k on X
Riemann Hypothesis	Trace formula

The contents of this survey paper are organized as follows,

I Local class field theory and the classification of injective factors.

In this section I shall first look back at my early work on the classification of von Neumann algebras and cast it in the unusual light of André Weil's Basic Number Theory ([W1]). It then appears as a clear substitute for the missing Brauer theory of central simple algebras for Archimedian local fields.

Let K be a *local* field, i.e. a nondiscrete locally compact field. The action of $K^* = GL_1(K)$ on the additive group K by multiplication,

$$(1) \qquad (\lambda, x) \to \lambda x \qquad \forall \lambda \in K^*, \ x \in K,$$

together with the uniqueness, up to scale, of the haar measure of the additive group K, yield a homomorphism,

$$(2) \qquad a \in K^* \to |a| \in \mathbb{R}_+^*,$$

from K^* to $\mathbb{R}_+^*$, called the *module* of K. Its range

$$(3) \qquad \mathrm{Mod}(K) = \{|\lambda| \in \mathbb{R}_+^* \ ; \ \lambda \in K^*\}$$

is a closed subgroup of $\mathbb{R}_+^*$.

The fields $\mathbb{R}$, $\mathbb{C}$ and $\mathbb{H}$ (of quaternions) are the only one with $\mathrm{Mod}(K) = \mathbb{R}_+^*$, they are called Archimedian local fields.

Let K be a non Archimedian local field, then

$$(4) \qquad R = \{x \in K \ ; \ |x| \leq 1\},$$

is the unique maximal compact subring of K and the quotient R/P of R by its unique maximal ideal is a finite field $\mathbb{F}_q$ (with $q = p^\ell$ a prime power). One has,

$$(5) \qquad\qquad \mathrm{Mod}(K) = q^{\mathbb{Z}} \subset \mathbb{R}_+^* \, .$$

Let K be commutative. An extension $K \subset K'$ of finite degree of K is called *unramified* iff the dimension of K' over K is the order of $\mathrm{Mod}(K')$ as a subgroup of $\mathrm{Mod}(K)$. When this is so, the field K' is commutative, is generated over K by roots of unity of order prime to q, and is a cyclic Galois extension of K with Galois group generated by the automorphism $\theta \in \mathrm{Aut}_K(K')$ such that,

$$(6) \qquad\qquad \theta(\mu) = \mu^q \, ,$$

for any root of unity of order prime to q in K'.

The unramified extensions of finite degree of K are classified by the subgroups,

$$(7) \qquad\qquad \Gamma \subset \mathrm{Mod}(K) \, , \ \Gamma \neq \{1\} \, .$$

Let then $\overline{K}$ be an algebraic closure of K, $K_{\mathrm{sep}} \subset \overline{K}$ the separable algebraic closure, $K_{\mathrm{ab}} \subset K_{\mathrm{sep}}$ the maximal abelian extension of K and $K_{\mathrm{un}} \subset K_{\mathrm{ab}}$ the maximal unramified extension of K, i.e. the union of all unramified extensions of finite degree. One has,

$$(8) \qquad\qquad K \subset K_{\mathrm{un}} \subset K_{\mathrm{ab}} \subset K_{\mathrm{sep}} \subset \overline{K} \, ,$$

and the Galois group $\mathrm{Gal}(K_{\mathrm{un}} : K)$ is topologically generated by θ called the Frobenius automorphism.

The correspondence (7) is given by,

$$(9) \qquad\qquad K' = \{x \in K_{\mathrm{un}} \, ; \ \theta_\lambda(x) = x \ \ \forall \lambda \in \Gamma\} \, ,$$

with rather obvious notations so that θ_q is the θ of (6). Let then W_K be the subgroup of $\mathrm{Gal}(K_{\mathrm{ab}} : K)$ whose elements induce on K_{un} an integral power of the Frobenius automorphism. One endows W_K with the locally compact topology dictated by the exact sequence of groups,

$$(10) \qquad\qquad 1 \to \mathrm{Gal}(K_{\mathrm{ab}} : K_{\mathrm{un}}) \to W_K \to \mathrm{Mod}(K) \to 1 \, ,$$

and the main result of local class field theory asserts the existence of a canonical isomorphism,

$$(11) \qquad\qquad W_K \overset{\sim}{\to} K^* \, ,$$

compatible with the module.

The basic step in the construction of the isomorphism (11) is the classification of finite dimensional central simple algebras A over K. Any such algebra is of the form,

$$(12) \qquad\qquad A = M_n(D) \, ,$$

where D is a (central) division algebra over K and the symbol M_n stands for $n \times n$ matrices.

Moreover D is the crossed product of an unramified extension K' of K by a 2-cocycle on its cyclic Galois group. Elementary group cohomology then yields the isomorphism,

$$(13) \qquad \qquad \mathrm{Br}(K) \overset{\eta}{\to} \mathbb{Q}/\mathbb{Z} \,,$$

of the Brauer group of classes of central simple algebras over K (with tensor product as the group law), with the group $\mathbb{Q}/\mathbb{Z}$ of roots of 1 in $\mathbb{C}$.

All the above discussion was under the assumption that K is non Archimedian. For Archimedian fields $\mathbb{R}$ and $\mathbb{C}$ the same questions have an idiotically simple answer. Since $\mathbb{C}$ is algebraically closed one has $K = \overline{K}$ and the whole picture collapses. For $K = \mathbb{R}$ the only non trivial value of the Hasse invariant η is

$$(14) \qquad \qquad \eta(\mathbb{H}) = -1 \,.$$

A Galois group G is by construction totally disconnected so that a morphism from K^* to G is necessarily trivial on the connected component of $1 \in K^*$.

Let k be a *global* field, i.e. a discrete cocompact subfield of a (non discrete) locally compact semi-simple commutative ring A. (Cf. Iwasawa *Ann. of Math.* **57** (1953).) The topological ring A is canonically associated to k and called the Adele ring of k, one has,

$$(15) \qquad \qquad A = \prod_{\mathrm{res}} k_v \,,$$

where the product is the restricted product of the local fields k_v labelled by the places of k.

When the characteristic of k is $p > 1$ so that k is a function field over $\mathbb{F}_q$, one has

$$(16) \qquad \qquad k \subset k_{\mathrm{un}} \subset k_{\mathrm{ab}} \subset k_{\mathrm{sep}} \subset \overline{k} \,,$$

where, as above $\overline{k}$ is an algebraic closure of k, k_{sep} the separable algebraic closure, k_{ab} the maximal abelian extension and k_{un} is obtained by adjoining to k all roots of unity of order prime to p.

One defines the Weil group W_k as above as the subgroup of $\mathrm{Gal}(k_{\mathrm{ab}} : k)$ of those automorphisms which induce on k_{un} an integral power of θ,

$$(17) \qquad \qquad \theta(\mu) = \mu^q \qquad \forall \, \mu \text{ root of 1 of order prime to } p \,.$$

The main theorem of global class field theory asserts the existence of a canonical isomorphism,

$$(18) \qquad \qquad W_k \simeq C_k = GL_1(A)/GL_1(k) \,,$$

of locally compact groups.

When k is of characteristic 0, i.e. is a number field, one has a canonical isomorphism,

$$(19) \qquad\qquad \mathrm{Gal}(k_{\mathrm{ab}} : k) \simeq C_k/D_k \,,$$

where D_k is the connected component of identity in the Idele class group $C_k = GL_1(A)/GL_1(k)$, but because of the Archimedian places of k there is no interpretation of C_k analogous to the Galois group interpretation for function fields. According to A. Weil ([W4]),

"La recherche d'une interprétation pour C_k si k est un corps de nombres, analogue en quelque manière à l'interprétation par un groupe de Galois quand k est un corps de fonctions, me semble constituer l'un des problèmes fondamentaux de la théorie des nombres à l'heure actuelle ; il se peut qu'une telle interprétation renferme la clef de l'hypothèse de Riemann ...".

Galois groups are by construction projective limits of the finite groups attached to finite extensions. To get connected groups one clearly needs to relax this finiteness condition which is the same as the finite dimensionality of the central simple algebras. Since Archimedian places of k are responsible for the non triviality of D_k it is natural to ask the following preliminary question,

"Is there a non trivial Brauer theory of central simple algebras over $\mathbb{C}$."

As we shall see shortly the *approximately finite dimensional* simple central algebras over $\mathbb{C}$ provide a satisfactory answer to this question. They are classified by their module,

$$(20) \qquad\qquad \mathrm{Mod}(M) \underset{\sim}{\subset} \mathbb{R}_+^* \,,$$

which is a virtual closed subgroup of $\mathbb{R}_+^*$.

Let us now explain this statement with more care. First we exclude the trivial case $M = M_n(\mathbb{C})$ of matrix algebras. Next $\mathrm{Mod}(M)$ is a virtual subgroup of $\mathbb{R}_+^*$, in the sense of G. Mackey, i.e. an ergodic action of $\mathbb{R}_+^*$. All ergodic flows appear and M_1 is isomorphic to M_2 iff $\mathrm{Mod}(M_1) \cong \mathrm{Mod}(M_2)$.

The birth place of central simple algebras is as the commutant of isotypic representations. When one works over $\mathbb{C}$ it is natural to consider unitary representations in Hilbert space so that we shall restrict our attention to algebras M which appear as commutants of unitary representations. They are called von Neumann algebras. The terms central and simple keep their usual algebraic meaning.

The classification involves three independent parts,

(A) The definition of the invariant $\mathrm{Mod}(M)$ for arbitrary factors (central von Neumann algebras).

(B) The equivalence of all possible notions of approximate finite dimensionality.

(C) The proof that Mod is a complete invariant and that all virtual subgroups are obtained.

The module of a factor M was first defined ([Co2]) as a closed subgroup of $\mathbb{R}_+^*$ by the equality

$$(21) \qquad S(M) = \bigcap_\varphi \; \mathrm{Spec}(\Delta_\varphi) \subset \mathbb{R}_+$$

where φ varies among (faithful, normal) states on M, i.e. linear forms $\varphi : M \to \mathbb{C}$ such that,

$$(22) \qquad \varphi(x^*x) \geq 0 \qquad \forall\, x \in M \,, \; \varphi(1) = 1 \,,$$

while the operator Δ_φ is the *modular operator* ([T])

$$(23) \qquad \Delta_\varphi = S_\varphi^* \, S_\varphi \,,$$

which is the *module* of the involution $x \to x^*$ in the Hilbert space attached to the sesquilinear form,

$$(24) \qquad \langle x,y \rangle = \varphi(y^*x) \,, \; x,y \in M \,.$$

In the case of local fields the module was a group homomorphism ((2)) from K^* to $\mathbb{R}_+^*$. The counterpart for factors is the group homomorphism, ([Co2])

$$(25) \qquad \delta : \mathbb{R} \to \mathrm{Out}(M) = \mathrm{Aut}(M)/\mathrm{Int}(M) \,,$$

from the additive group $\mathbb{R}$ viewed as the dual of $\mathbb{R}_+^*$ for the pairing,

$$(26) \qquad (\lambda, t) \to \lambda^{it} \qquad \forall\, \lambda \in \mathbb{R}_+^* \,, \; t \in \mathbb{R} \,,$$

to the group of automorphism classes of M modulo inner automorphisms.

The virtual subgroup,

$$(27) \qquad \mathrm{Mod}(M) \underset{\sim}{\subset} \mathbb{R}_+^* \,,$$

is the *flow of weights* ([Ta],[K],[CT]) of M. It is obtained from the module δ as the dual action of $\mathbb{R}_+^*$ on the abelian algebra,

$$(28) \qquad C = \text{Center of } M \rtimes_\delta \mathbb{R} \,,$$

where $M \rtimes_\delta \mathbb{R}$ is the crossed product of M by the modular automorphism group δ.

This takes care of (A), to describe (B) let us simply state the equivalence ([Co1]) of the following conditions

$$(29) \qquad \begin{array}{l} M \text{ is the closure of the union of an increasing sequence of} \\ \text{finite dimensional algebras.} \end{array}$$

$$(30) \qquad \begin{array}{l} M \text{ is complemented as a subspace of the normed space of} \\ \text{all operators in a Hilbert space.} \end{array}$$

The condition (29) is obviously what one would expect for an approximately finite dimensional algebra. Condition (30) is similar to *amenability* for discrete groups and the implication (30) $\Rightarrow$ (29) is a very powerful tool.

We refer to [Co1],[K],[Ha] for (C) and we just describe the actual construction of the central simple algebra M associated to a given virtual subgroup,

$$(31) \qquad\qquad \underset{\sim}{\Gamma} \subseteq \mathbb{R}_+^* .$$

Among the approximately finite dimensional factors (central von Neumann algebras), only two are not simple. The first is the algebra

$$(32) \qquad\qquad M_\infty(\mathbb{C}) ,$$

of all operators in Hilbert space. The second factor is the unique approximately finite dimensional factor of type II_∞. It is

$$(33) \qquad\qquad R_{0,1} = R \otimes M_\infty(\mathbb{C}) ,$$

where R is the unique approximately finite dimensional factor with a finite trace τ_0, i.e. a state such that,

$$(34) \qquad\qquad \tau_0(xy) = \tau_0(yx) \qquad \forall\, x, y \in R .$$

The tensor product of τ_0 by the standard semifinite trace on $M_\infty(\mathbb{C})$ yields a semifinite trace τ on $R_{0,1}$. There exists, up to conjugacy, a unique one parameter group of automorphisms $\theta_\lambda \in \mathrm{Aut}(R_{0,1})$, $\lambda \in \mathbb{R}_+^*$ such that,

$$(35) \qquad\qquad \tau(\theta_\lambda(a)) = \lambda\tau(a) \qquad \forall\, a \in \mathrm{Domain}\,\tau\, , \ \lambda \in \mathbb{R}_+^* .$$

Let first $\Gamma \subset \mathbb{R}_+^*$ be an ordinary closed subgroup of $\mathbb{R}_+^*$. Then the corresponding factor R_Γ with module Γ is given by the equality:

$$(36) \qquad\qquad R_\Gamma = \{x \in R_{0,1} \,;\, \theta_\lambda(x) = x \quad \forall\, \lambda \in \Gamma\} ,$$

in perfect analogy with (9).

A virtual subgroup $\underset{\sim}{\Gamma} \subset \mathbb{R}_+^*$ is by definition an ergodic action of $\mathbb{R}_+^*$ on an abelian von Neumann algebra A, and the formula (36) easily extends to,

$$(37) \qquad\qquad R_\Gamma = \{x \in R_{0,1} \otimes A \,;\, (\theta_\lambda \otimes \alpha_\lambda)\, x = x \quad \forall\, \lambda \in \mathbb{R}_+^*\} .$$

(This reduces to (36) for the action of $\mathbb{R}_+^*$ on the algebra $A = L^\infty(X)$ where X is the homogeneous space $X = \mathbb{R}_+^*/\Gamma$.)

The pair $(R_{0,1}, \theta_\lambda)$ arises very naturally in geometry from the geodesic flow of a compact Riemann surface (of genus > 1). Let $V = S^*\Sigma$ be the unit cosphere bundle of such a surface Σ, and F be the stable foliation of the geodesic flow. The latter defines a one parameter group of automorphisms of the foliated manifold (V, F)

---

and thus a one parameter group of automorphisms θ_λ of the von Neumann algebra $L^\infty(V,F)$.

This algebra is easy to describe, its elements are random operators $T = (T_f)$, i.e. bounded measurable families of operators T_f parametrized by the leaves f of the foliation. For each leaf f the operator T_f acts in the Hilbert space $L^2(f)$ of square integrable densities on the manifold f. Two random operators are identified if they are equal for almost all leaves f (i.e. a set of leaves whose union in V is negligible). The algebraic operators of sum and product are given by,

$$(38) \qquad (T_1 + T_2)_f = (T_1)_f + (T_2)_f, \ (T_1 T_2)_f = (T_1)_f (T_2)_f,$$

i.e. are effected pointwise.

One proves that,

$$(39) \qquad L^\infty(V,F) \simeq R_{0,1},$$

and that the geodesic flow θ_λ satisfies (35). Indeed the foliation (V,F) admits up to scale a unique transverse measure Λ and the trace τ is given (cf. [C]) by the formal expression,

$$(40) \qquad \tau(T) = \int \mathrm{Trace}(T_f)\, d\Lambda(f),$$

since the geodesic flow satisfies $\theta_\lambda(\Lambda) = \lambda\Lambda$ one obtains (35) from simple geometric considerations. The formula (37) shows that most approximately finite dimensional factors already arise from foliations, for instance the unique approximately finite dimensional factor R_∞ such that,

$$(41) \qquad \mathrm{Mod}(R_\infty) = \mathbb{R}_+^*,$$

arises from the codimension 1 foliation of $V = S^*\Sigma$ generated by F and the geodesic flow.

In fact this relation between the classification of central simple algebras over $\mathbb{C}$ and the geometry of foliations goes much deeper. For instance using cyclic cohomology together with the following simple fact,

$$(42) \qquad \text{``A connected group can only act trivially on a homotopy invariant cohomology theory''},$$

one proves (cf. [C]) that for any codimension are foliation F of a compact manifold V with non vanishing Godbillon-Vey class one has,

$$(43) \qquad \mathrm{Mod}(M) \text{ has finite covolume in } \mathbb{R}_+^*,$$

where $M = L^\infty(V,F)$ and a virtual subgroup of finite covolume is a flow with a finite invariant measure.

II Global class field theory and spontaneous symmetry breaking.

In the above discussion of approximately finite dimensional central simple algebras, we have been working locally over $\mathbb{C}$. We shall now describe a particularly interesting example ([BC]) of Hecke algebra intimately related to arithmetic, and defined over $\mathbb{Q}$.

Let $\Gamma_0 \subset \Gamma$ be an almost normal subgroup of a discrete group Γ, i.e. one assumes,

(1) $\Gamma_0 \cap s\,\Gamma_0\,s^{-1}$ has finite index in Γ_0 $\forall\, s \in \Gamma$.

Equivalently the orbits of the left action of Γ_0 on Γ/Γ_0 are all finite. One defines the Hecke algebra,

(2) $\mathcal{H}(\Gamma, \Gamma_0)\,,$

as the convolution algebra of integer valued Γ_0 biinvariant functions with finite support. For any field k one lets,

(3) $\mathcal{H}_k(\Gamma, \Gamma_0) = \mathcal{H}(\Gamma, \Gamma_0) \otimes_{\mathbb{Z}} k\,,$

be obtained by extending the coefficient ring from $\mathbb{Z}$ to k. We let $\Gamma = P_{\mathbb{Q}}^+$ be the group of 2×2 rational matrices,

(4) $\Gamma = \left\{ \begin{bmatrix} 1 & b \\ 0 & a \end{bmatrix} ;\ a \in \mathbb{Q}^+ ,\ b \in \mathbb{Q} \right\} ,$

and $\Gamma_0 = P_{\mathbb{Z}}^+$ be the subgroup of integral matrices,

(5) $\Gamma_0 = \left\{ \begin{bmatrix} 1 & n \\ 0 & 1 \end{bmatrix} ;\ n \in \mathbb{Z} \right\} .$

One checks that Γ_0 is almost normal in Γ.

To obtain a central simple algebra over $\mathbb{C}$ in the sense of the previous section we just take the commutant of the right regular representation of Γ on $\Gamma_0 \backslash \Gamma$, i.e. the weak closure of $\mathcal{H}_{\mathbb{C}}(\Gamma, \Gamma_0)$ in the Hilbert space,

(6) $\ell^2(\Gamma_0 \backslash \Gamma)\,,$

of Γ_0 left invariant function on Γ with norm square,

$$\|\xi\|^2 = \sum_{\gamma \in \Gamma_0 \backslash \Gamma} |\xi(\gamma)|^2 . \tag{7}$$

This central simple algebra over $\mathbb{C}$ is approximately finite dimensional and its module is $\mathbb{R}_+^*$ so that it is the same as R_∞ of (I.41).

In particular its modular automorphism group is highly non trivial and one can compute it explicitly for the state φ associated to the vector $\xi_0 \in \ell^2(\Gamma_0 \backslash \Gamma)$ corresponding to the left coset Γ_0.

The modular automorphism group σ_t^φ leaves the dense subalgebra $\mathcal{H}_{\mathbb{C}}(\Gamma, \Gamma_0) \subset R_\infty$ globally invariant and is given by the formula,

$$\sigma_t^\varphi(f)(\gamma) = L(\gamma)^{-it} R(\gamma)^{it} f(\gamma) \qquad \forall \gamma \in \Gamma_0 \backslash \Gamma / \Gamma_0 \tag{8}$$

for any $f \in \mathcal{H}_{\mathbb{C}}(\Gamma, \Gamma_0)$. Here we let,

$$\begin{aligned} L(\gamma) &= \text{ Cardinality of the image of } \Gamma_0 \gamma \Gamma_0 \text{ in } \Gamma / \Gamma_0 \\ R(\gamma) &= \text{ Cardinality of the image of } \Gamma_0 \gamma \Gamma_0 \text{ in } \Gamma_0 \backslash \Gamma . \end{aligned} \tag{9}$$

This is enough to make contact with the formalism of quantum statistical mechanics which we now briefly describe.

A quantum statistical system is given by,

1) The C^* algebra of observables A,

2) The time evolution $(\sigma_t)_{t \in \mathbb{R}}$ which is a one parameter group of automorphisms of A.

An equilibrium or KMS (for Kubo-Martin and Schwinger) state, at inverse temperature β is a state φ on A which fulfills the following condition,

(10) For any $x, y \in A$ there exists a bounded holomorphic function (continuous on the closed strip), $F_{x,y}(z)$, $0 \le \operatorname{Im} z \le \beta$ such that

$$\begin{aligned} F_{x,y}(t) &= \varphi(x\,\sigma_t(y)) & \forall t \in \mathbb{R} \\ F_{x,y}(t+i\beta) &= \varphi(\sigma_t(y)x) & \forall t \in \mathbb{R} . \end{aligned}$$

For fixed β the KMS$_\beta$ states form a Choquet simplex and thus decompose uniquely as a statistical superposition from the pure phases given by the extreme points. For interesting systems with nontrivial interaction, one expects in general that for large temperature T, (i.e. small β since $\beta = \frac{1}{T}$ up to a conversion factor) the disorder will be predominant so that there will exist only one KMS$_\beta$ state. For low enough temperatures some order should set in and allow for the coexistence of distinct thermodynamical phases so that the simplex K_β of KMS$_\beta$ states should be non trivial. A given symmetry group G of the system will necessarily act trivially on K_β for large T since K_β is a point, but acts in general non trivially on K_β for small T so that it is no longer a symmetry of a given pure phase. This phenomenon of *spontaneous symmetry breaking* as well as the very particular properties of the critical temperature T_c at the boundary of the two regions are corner stones of statistical mechanics.

In our case we just let A be the C^* algebra which is the *norm* closure of $\mathcal{H}_C(\Gamma, \Gamma_0)$ in the algebra of operators in $\ell^2(\Gamma_0 \backslash \Gamma)$. We let $\sigma_t \in \mathrm{Aut}(A)$ be the unique extension of the automorphisms σ_t^φ of (8).

For $\beta = 1$ it is tautological that φ is a KMS$_\beta$ state since we obtained σ_t^φ precisely this way ([T]). One proves ([BC]) that for any $\beta \leq 1$ (i.e. for $T = 1$) there exists one and only one KMS$_\beta$ state.

The compact group G,

$$(11) \qquad\qquad G = C_{\mathbb{Q}}/D_{\mathbb{Q}},$$

quotient of the Idele class group $C_{\mathbb{Q}}$ by the connected component of identity $D_{\mathbb{Q}} \simeq \mathbb{R}_+^*$, acts in a very simple and natural manner as symmetries of the system (A, σ_t). (To see this one notes that the right action of Γ on $\Gamma_0 \backslash \Gamma$ extends to the action of $P_{\mathcal{A}}$ on the restricted product of the trees of $SL(2, \mathbb{Q}_p)$ where $\mathcal{A}$ is the ring of finite Adeles (cf. [BC]).

For $\beta > 1$ this symmetry group G of our system, is spontaneously broken, the compact convex sets K_β are non trivial and have the same structure as K_∞, which we now describe. First some terminology, a KMS$_\beta$ state for $\beta = \infty$ is called a *ground state* and the KMS$_\infty$ condition is equivalent to *positivity of energy* in the corresponding Hilbert space representation.

Remember that $\mathcal{H}_C(\Gamma, \Gamma_0)$ contains $\mathcal{H}_{\mathbb{Q}}(\Gamma, \Gamma_0)$ so,

$$(12) \qquad\qquad \mathcal{H}_{\mathbb{Q}}(\Gamma, \Gamma_0) \subset A.$$

By [BC] theorem 5 and proposition 24 one has,

Theorem. *Let $\mathcal{E}(K_\infty)$ be the set of extremal KMS$_\infty$ states.*

a) *The group G acts freely and transitively on $\mathcal{E}(K_\infty)$ by composition, $\varphi \to \varphi \circ g^{-1}$, $\forall g \in G$.*

b) *For any $\varphi \in \mathcal{E}(K_\infty)$ one has,*

$$\varphi(\mathcal{H}_{\mathbb{Q}}) = \mathbb{Q}_{\mathrm{ab}},$$

and for any element $\alpha \in \mathrm{Gal}(\mathbb{Q}_{\mathrm{ab}} : \mathbb{Q})$ there exists a unique extension of $\alpha \circ \varphi$, by continuity, as a state of A. One has $\alpha \circ \varphi \in \mathcal{E}(K_\infty)$.

c) *The map $\alpha \to \varphi^{-1}(\alpha \circ \varphi) \in G = C_k/D_k$ defined for $\alpha \in \mathrm{Gal}(\mathbb{Q}_{\mathrm{ab}} : \mathbb{Q})$ is the isomorphism of global class field theory (I.19).*

This last map is independent of the choice of φ. What is quite remarkable in this result is that the existence of the subalgebra $\mathcal{H}_{\mathbb{Q}} \subset \mathcal{H}_C$ allows to bring into action the Galois group of $\mathbb{C}$ on the *values of states*. Since the Galois group of $\mathbb{C} : \mathbb{Q}$ is (except for $z \to \bar{z}$) formed of *discontinuous* automorphisms it is quite surprising that its action can actually be compatible with the characteristic *positivity* of states.

It is by no means clear how to extend the above construction to arbitrary number fields k while preserving the 3 results of the theorem. The ideas of G. Moore ([Mo]) could well be relevant. There is however an easy computation which relates the above construction to an object which makes sense for any global field k. Indeed if we let as above R_∞ be the weak closure of $\mathcal{H}_C(\Gamma, \Gamma_0)$ in $\ell^2(\Gamma_0 \backslash \Gamma)$, we can compute the associated pair $(R_{0,1}, \theta_\lambda)$ of section I.

By the result of [Laca] the C^* algebra closure of $\mathcal{H}_C$ is a full corner of the crossed product C^* algebra,

$$(13) \qquad\qquad C_0(\mathcal{A}) \rtimes \mathbb{Q}_+^* ,$$

where $\mathcal{A}$ is the locally compact space of finite Adeles. It follows immediately that,

$$(14) \qquad\qquad R_{0,1} = L^\infty(\mathbb{Q}_A) \rtimes \mathbb{Q}^* ,$$

i.e. the von Neumann algebra crossed product of the L^∞ functions on Adeles of $\mathbb{Q}$ by the action of $\mathbb{Q}^*$ by multiplication.

The one parameter group of automorphisms, $\theta_\lambda \in \mathrm{Aut}(R_{0,1})$, is obtained as the restriction to,

$$(15) \qquad\qquad D_\mathbb{Q} = \mathbb{R}_+^* ,$$

of the obvious action of the Idele class group $C_\mathbb{Q}$,

$$(16) \qquad\qquad (g, x) \to g\,x \qquad \forall\, g \in C_\mathbb{Q},\ x \in A_\mathbb{Q}/\mathbb{Q}^* ,$$

on the space $X = A_\mathbb{Q}/\mathbb{Q}^*$ of Adele classes.

Our next goal will be to show that the latter space is intimately related to the *zeros* of the Hecke L-functions with Grössencharakter.

(We showed in [BC] that the partition function of the above system is the Riemann zeta function.)

III Zeros of zeta and random matrices.

It is an old idea, due to Polya and Hilbert that in order to understand the location of the zeros of the Riemann zeta function, one should find a Hilbert space $\mathcal{H}$ and an operator D in $\mathcal{H}$ whose spectrum is given by the non trivial zeros of the zeta function. The hope then is that suitable selfadjointness properties of D (of $i\left(D - \frac{1}{2}\right)$ more precisely) or positivity properties of $\Delta = D(1 - D)$ will be easier to handle than the original conjecture. The main reasons why this idea should be taken seriously are first the work of A. Selberg ([Se]) in which a suitable Laplacian Δ is related in the above way to an analogue of the zeta function, and secondly the theoretical ([M][B][KS]) and experimental evidence ([O][BG]) on the fluctuations of the spacing between consecutive zeros of zeta. The number of zeros of zeta whose imaginary part is less than $E > 0$,

(1) $$N(E) = \# \text{ of zeros } \rho \, , \; 0 < \operatorname{Im} \rho < E$$

has an asymptotic expression ([R]) given by

(2) $$N(E) = \frac{E}{2\pi}\left(\log\left(\frac{E}{2\pi}\right) - 1\right) + \frac{7}{8} + o(1) + N_{\mathrm{osc}}(E)$$

where the oscillatory part of this step function is

(3) $$N_{\mathrm{osc}}(E) = \frac{1}{\pi} \operatorname{Im} \log \zeta \left(\frac{1}{2} + iE\right)$$

assuming that E is not the imaginary part of a zero and taking for the logarithm the branch which is 0 at $+\infty$ (connected to $\frac{1}{2} + iE$ by a straight horizontal line).

One shows (cf. [Pat]) that $N_{\mathrm{osc}}(E)$ is $O(\log E)$. In the decomposition (2) the two terms $\langle N(E)\rangle = N(E) - N_{\mathrm{osc}}(E)$ and $N_{\mathrm{osc}}(E)$ play an independent role. The first one $\langle N(E)\rangle$ which gives the average density of zeros is computed as follows. Let $\xi(s) = \pi^{-s/2}\Gamma(s/2)\zeta(s)$. Then $\xi(s) = \xi(1 - s)$ and $\xi(\bar{s}) = \overline{\xi(s)}$ so that ξ is real on the line $\operatorname{Re}(s) = 1/2$ and $\operatorname{Im}\log\xi(s) = \operatorname{Im}(-s/2\log\pi + \log\Gamma(s/2) + \log\zeta(s)) \in \mathbb{Z}\pi$. This shows that,

$$N(E) = 1 + \frac{1}{\pi}\left(-\frac{E}{2}\log\pi + \operatorname{Im}\log\Gamma(\frac{1}{4} + i\frac{E}{2})\right) + \frac{1}{\pi}\operatorname{Im}\log\zeta(\frac{1}{2} + iE)$$

The asymptotic expansion of $\langle N(E)\rangle$ is thus given by the Stirling formula

$$\Gamma(z) = e^{-z}z^{z-1/2}\sqrt{2\pi}(1 + O(1/z))$$

valid for $-\pi < arg z < \pi$. Hence the global behavior of the function $N(E)$ is described by the smooth part

$$< N(E) >= \frac{E}{2\pi}\left(\log \frac{E}{2\pi} - 1\right) + \frac{7}{8} + o(1)$$

The second $N_{\text{osc}}(E)$ is a manifestation of the randomness of the actual location of the zeros, and to eliminate the role of the density one returns to the situation of uniform density by the transformation

(4) $\qquad x_j = \langle N(E_j)\rangle \quad (E_j \text{ the } j^{\text{th}} \text{ imaginary part of zero of zeta}).$

Thus the spacing between two consecutive x_j is now 1 in average and the only information that remains is in the statistical fluctuation. As it turns out ([M][O]) these fluctuations are the same as the fluctuations of the eigenvalues of a random hermitian matrix of very large size.

H. Montgomery [M] proved (assuming RH) a weakening of the following conjecture (with $\alpha, \beta > 0$),

$$\text{Card}\{(i,j)\,;\,i,j \in 1,\ldots,M\,;\,x_i - x_j \in [\alpha,\beta]\}$$

(5) $$\sim M \int_\alpha^\beta \left(1 - \left(\frac{\sin(\pi u)}{\pi u}\right)^2\right) du$$

This law (5) is precisely the same as the correlation between eigenvalues of hermitian matrices of the Gaussian Unitary Ensemble ([Me]) which we shall now describe. Moreover, numerical tests due to A. Odlyzko ([O][BG]) have confirmed with great precision the behaviour (5) as well as the analogous behaviour for more than two zeros. In [KS], N. Katz and P. Sarnak proved an analogue of the Montgomery-Odlyzko law for zeta and L-functions of function fields over curves.

The Gaussian Unitary Ensemble is the probability measure on the vector space $M_N(\mathbb{C})_{sa}$ of $N \times N$ Hermitian matrices, given, up to normalization, by the density $e^{-Trace A^2} dA$, $A \in M_N(\mathbb{C})_{sa}$, where dA denotes the Haar measure with respect to the additive structure. This measure is unitarily invariant.

We let $p(E_1, E_2, \ldots E_N)dE_1 \ldots dE_N$ be the corresponding measure on the eigenvalues $E_1, E_2, \ldots E_N$ of the matrix, i.e. the probability measure induced on the collections of eigenvalues by the Gaussian measure.

One has ([Me]),

(6) $$p(E_1, \ldots, E_N) = 1/N!\left(\det_{\substack{0 \le k \le N-1 \\ 1 \le l \le N}} (\Phi_k(E_l))\right)^2,$$

where Φ_n denotes the n-th Hermite function,

(7) $$\Phi_n(x) = \frac{(-1)^n \pi^{-1/4}}{\sqrt{2^n n!}}(\partial^n e^{-x^2})e^{x^2/2}.$$

One has $\Phi_n(x) = P_n(x)e^{-x^2/2}$, where P_n is the Hermite polynomial of degree n. The probability density of eigenvalues is by definition,

$$(8) \qquad R_N(E) = \int p(E, E_2, \ldots E_N) dE_2 \ldots dE_N.$$

Using (6) it is given, up to normalization, by the following expression,

$$\sum_{\sigma, \pi \in S_N} \epsilon(\sigma)\epsilon(\pi) \int \Phi_{\sigma(1)}(E)\Phi_{\sigma(2)}(E_2) \ldots \Phi_{\sigma(N)}(E_N)$$

$$\Phi_{\pi(1)}(E)\Phi_{\pi(2)}(E_2) \ldots \Phi_{\pi(N)}(E_N) dE_2 \ldots dE_N$$

The integral is nonzero only if $\sigma(k) = \pi(k)$, $k = 2, \ldots, N$, and hence only for $\sigma = \pi$. Since $\sigma(1) = \pi(1)$ can be any number between 0 and $N - 1$, one gets $R_N = \frac{1}{N} K_N(E)$, where

$$(9) \qquad K_N(E) = \sum_{j=0}^{N-1} \Phi_j(E)^2.$$

One has $K_N(E) = K_N(E, E)$, where $K_N = \sum_{j=0}^{N-1} |\Phi_j ><\Phi_j|$ is the orthogonal projection on the spectral subspace $H \leq 2N - 1$ of the harmonic oscillator $H = -\partial^2 + x^2 = p^2 + q^2$.

The asymptotic behaviour of $K_N(E)$ is then given by the semiclassical approximation. To the part of the spectrum $H \leq 2N - 1$ corresponds the disk of radius $\sqrt{2N - 1}$ in the (p, q) plane with measure $\frac{1}{2\pi} dp \wedge dq$. Then the asymptotic behavior of $\int f(E) K_N(E) dE = \text{Trace}(f(q)K_N)$, $N \to \infty$, is given by,

$$(10) \qquad \frac{1}{2\pi} \int_{p^2+q^2 \leq 2N} f(q) dp \wedge dq = \frac{1}{\pi} \int_{-\sqrt{2N}}^{\sqrt{2N}} f(E)\sqrt{2N - E^2} dE$$

and for R_N one gets the asymptotic behavior

$$(11) \qquad R_N(E) \sim \frac{1}{\pi N} \sqrt{2N - E^2}$$

i.e. the semicircle law.

This gives the density of eigenvalues and the analogue of the transformation (4) near $E = 0$ is $x_j = \frac{\sqrt{2N}}{\pi} E_j$. To study the local fluctuations one considers the two point correlation function,

$$(12) \qquad R_N(E_1, E_2) = \int p(E_1, E_2, E_3 \ldots E_N) dE_3 \ldots dE_N$$

and its limit behavior when $N \to \infty$, with $E_j = \frac{\pi x_j}{\sqrt{2N}}$, $j = 1, 2$. As before, one needs to compute,

$$\sum_{\sigma, \pi \in S_N} \epsilon(\sigma)\epsilon(\pi) \int \Phi_{\sigma(1)}(E)\Phi_{\sigma(2)}(E_2) \ldots \Phi_{\sigma(N)}(E_N)$$

$$\Phi_{\pi(1)}(E)\Phi_{\pi(2)}(E_2) \ldots \Phi_{\pi(N)}(E_N) dE_3 \ldots dE_N$$

One gets nonzero terms only if $\sigma(k) = \pi(k)$, $k = 3, \ldots, N$. It means that possible nonzero terms are obtained when $\sigma(1) = \pi(1)$, $\sigma(2) = \pi(2)$, and in this case $\epsilon(\sigma) = \epsilon(\pi)$ and when $\sigma(1) = \pi(2)$, $\sigma(2) = \pi(1)$, and in this case $\epsilon(\sigma) = -\epsilon(\pi)$. The value of the integral is thus,

$$\frac{1}{N(N-1)} \sum_{k,l} \Phi_k(E_1)^2 \Phi_l(E_2)^2 - \Phi_k(E_1)\Phi_k(E_2)\Phi_l(E_1)\Phi_l(E_2) =$$

$$\frac{1}{N(N-1)} \begin{vmatrix} K_N(E_1, E_1) & K_N(E_1, E_2) \\ K_N(E_1, E_2) & K_N(E_2, E_2) \end{vmatrix}$$

The asymptotic behavior of $K_N(x, y)$ is obtained using ([Me]),

$$(13) \quad K_N(x, y) = \sum_{n=0}^{N-1} \Phi_n(x)\Phi_n(y) = \left(\frac{N}{2}\right)^{1/2} \frac{\Phi_N(x)\Phi_{N-1}(y) - \Phi_N(y)\Phi_{N-1}(x)}{x - y}$$

from which one easily gets ([Me]),

$$(14) \quad K_N\left(\frac{\pi x_1}{\sqrt{2N}}, \frac{\pi x_2}{\sqrt{2N}}\right) \sim \frac{\sqrt{2N}}{\pi} \frac{\sin \pi x_1 \cos \pi x_2 - \sin \pi x_2 \cos \pi x_1}{\pi(x_1 - x_2)} =$$

$$\frac{\sqrt{2N}}{\pi} \frac{\sin \pi(x_1 - x_2)}{\pi(x_1 - x_2)}$$

One then obtains,

$$(15) \qquad R_N(\frac{\pi x_1}{\sqrt{2N}}, \frac{\pi x_2}{\sqrt{2N}}) \sim \frac{2}{\pi^2 N}\left(1 - \left(\frac{\sin \pi(x_1 - x_2)}{\pi(x_1 - x_2)}\right)^2\right),$$

which is identical to the Montgomery-Odlyzko law.

This is thus an excellent motivation to try and find a natural pair $(\mathcal{H}, D)$ where naturality should mean for instance that one should not even have to define the zeta function, let alone its analytic continuation, in order to obtain the pair (in order for instance to avoid the joke of defining $\mathcal{H}$ as the ℓ^2 space built on the zeros of zeta).

IV Quantum chaos and the minus sign problem.

We shall first describe following [B] the direct atempt to construct the Polya-Hilbert space from quantization of a classical dynamical system. The original motivation for the theory of random matrices comes from quantum mechanics. In this theory the quantization of the classical dynamical system given by the phase space X and hamiltonian h gives rise to a Hilbert space $\mathcal{H}$ and a selfadjoint operator H whose spectrum is the essential physical observable of the system. For complicated systems the only useful information about this spectrum is that, while the average part of the counting function,

(1) $N(E) = \#$ eigenvalues of H in $[0, E]$

is computed by a semiclassical approximation mainly as a volume in phase space, the oscillatory part,

(2) $N_{osc}(E) = N(E) - \langle N(E) \rangle$

is the same as for a random matrix, governed by the statistic dictated by the symmetries of the system.

In the absence of a magnetic field, i.e. for a classical hamiltonian of the form,

(3) $h = \dfrac{1}{2m} p^2 + V(q)$

where V is a real-valued potential on configuration space, there is a natural symmetry of classical phase space, called time reversal symmetry,

(4) $T(p, q) = (-p, q)$

which preserves h, and entails that the correct ensemble on the random matrices is not the above GUE but rather the gaussian orthogonal ensemble: GOE. Thus the oscillatory part $N_{osc}(E)$ behaves in the same way as for a random *real symmetric* matrix.

Of course H is just a specific operator in $\mathcal{H}$ and, in order that it behaves *generically* it is necessary (cf. [B]) that the classical hamiltonian system (X, h) be *chaotic* with isolated *periodic orbits* whose unstability exponents (i.e. the logarithm of the eigenvalues of the Poincaré return map acting on the transverse space to the orbits) are different from 0.

One can then ([B]) write down an asymptotic semiclassical approximation to the oscillatory function $N_{\mathrm{osc}}(E)$

$$(5) \qquad N_{\mathrm{osc}}(E) = \frac{1}{\pi} \operatorname{Im} \int_0^\infty \operatorname{Trace}(H - (E + i\eta))^{-1} \, id\eta$$

using the stationary phase approximation of the corresponding functional integral. For a system whose configuration space is 2-dimensional, this gives ([B] (15)),

$$(6) \qquad N_{\mathrm{osc}}(E) \simeq \frac{1}{\pi} \sum_{\gamma_p} \sum_{m=1}^{\infty} \frac{1}{m} \frac{1}{2\mathrm{sh}\left(\frac{m\lambda_p}{2}\right)} \sin(S_{\mathrm{pm}}(E))$$

where the γ_p are the primitive periodic orbits, the label m corresponds to the number of traversals of this orbit, while the corresponding unstability exponents are $\pm\lambda_p$. The phase $S_{\mathrm{pm}}(E)$ is up to a constant equal to $m E T_\gamma^{\#}$ where $T_\gamma^{\#}$ is the period of the primitive orbit γ_p.

The formula (6) gives very precious information ([B]) on the hypothetical "Riemann flow" whose quantization should produce the Polya-Hilbert space. The point is that the Euler product formula for the zeta function yields (cf. [B]) a similar asymptotic formula for $N_{\mathrm{osc}}(E)$ (3),

$$(7) \qquad N_{\mathrm{osc}}(E) \simeq \frac{-1}{\pi} \sum_p \sum_{m=1}^{\infty} \frac{1}{m} \frac{1}{p^{m/2}} \sin\left(m E \log p\right).$$

Comparing (6) and (7) gives the following information,

(A) The periodic primitive orbits should be labelled by the prime numbers $p = 2, 3, 5, 7, \ldots$, their periods should be the $\log p$ and their unstability exponents $\lambda_p = \pm \log p$.

Moreover, since each orbit is only counted once, the Riemann flow should not possess the symmetry T of (4) whose effect would be to duplicate the count of orbits. This last point excludes in particular the geodesic flows since they have the time reversal symmetry T. Thus we get

(B) The Riemann flow cannot satisfy time reversal symmetry.

However there are two important mismatches (cf. [B]) between the two formulas (6) and (7). The first one is the overall *minus sign* in front of formula (7), the second one is that though $2\mathrm{sh}\left(\frac{m\lambda_p}{2}\right) \sim p^{m/2}$ when $m \to \infty$, we do not have an equality for finite values of m.

We shall see in the next section how to overcome this $-$ sign problem. To put the solution in physics terminology, the spectral interpretation will appear in a natural manner as an absorption spectrum. Recall that spectral lines which are observed in spectroscopy (e.g. from the light coming from distant stars) are of two kinds: on the one hand one observes emission lines which are bright lines on a dark background, on the other hand one observes absorption lines which are dark lines on a bright background. It is the latter which will serve as a model for our spectral realization of zeros of zeta.

V Spectral interpretation of critical zeros.

The very same $-$ sign appears in the Riemann-Weil explicit formula ([W3]) which we now briefly describe. One lets k be a global field. One identifies the quotient $C_k/C_{k,1}$ with the range of the module,

(1) $$N = \{|g| \, ; \, g \in C_k\} \subset \mathbb{R}_+^* \, .$$

One endows N with its normalized Haar measure $d^* x$ where for modulated groups the normalization is as in Weil ([W3]),

(2) $$\int_{1 \leq |u| \leq \Lambda} d^* u \sim \log \Lambda \qquad \text{for } \Lambda \to \infty \, ,$$

Given a function F on N such that, for some $b > \frac{1}{2}$,

$$|F(\nu)| = 0(\nu^b) \quad \nu \to 0 \, , \quad |F(\nu)| = 0(\nu^{-b}) \, , \quad \nu \to \infty \, ,$$

one lets,

(3) $$\Phi(s) = \int_N F(\nu) \, \nu^{1/2-s} \, d^* \nu \, .$$

Given a Grössencharakter $\mathcal{X}$, i.e. a character of C_k and any ρ in the strip $0 < \text{Re}(\rho) < 1$, one lets $N(\mathcal{X}, \rho)$ be the order of $L(\mathcal{X}, s)$ at $s = \rho$. One lets,

(4) $$S(\mathcal{X}, F) = \sum_\rho N(\mathcal{X}, \rho) \, \Phi(\rho)$$

where the sum takes place over ρ's in the above open strip. One then defines a distribution Δ on C_k by,

(5) $$\Delta = \log |d^{-1}| \, \delta_1 + D - \sum_v D_v \, ,$$

where δ_1 is the Dirac mass at $1 \in C_k$, where d is a differential idele of k so that $|d|^{-1}$ is up to sign the discriminant of k when char $(k) = 0$ and is q^{2g-2} when k is a function field over a curve of genus g with coefficients in the finite field $\mathbb{F}_q$.

The distribution D is given by,

(6) $$D(f) = \int_{C_k} f(w) \, (|w|^{1/2} + |w|^{-1/2}) \, d^* w$$

where the Haar measure d^*w is normalized. The distributions D_v are labeled by the places v of k and are obtained as follows. For each v one considers the natural proper homomorphism,

$$(7) \qquad\qquad k_v^* \to C_k \ , \ x \to \text{class of } (1, \ldots, x, 1 \ldots)$$

of the multiplicative group of the local field k_v in the idele class group C_k.

One then has,

$$(8) \qquad\qquad D_v(f) = Pfw \int_{k_v^*} \frac{f(u)}{|1 - u|} |u|^{1/2} d^*u$$

where the Haar measure d^*u is normalized, and where the Weil Principal value Pfw of the integral is obtained as follows, for a local field $K = k_v$,

$$(9) \qquad\qquad Pfw \int_{k_v^*} 1_{R_v^*} \frac{1}{|1 - u|} d^*u = 0 \,,$$

if the local field k_v is non Archimedean, and otherwise:

$$(10) \qquad\qquad Pfw \int_{k_v^*} \varphi(u) \, d^*u = PF_0 \int_{\mathbb{R}_+^*} \psi(\nu) \, d^*\nu \,,$$

where $\psi(\nu) = \int_{|u|=\nu} \varphi(u) \, d_\nu u$ is obtained by integrating φ over the fibers, while

$$(11) \qquad PF_0 \int \psi(\nu) d^*\nu = 2\log(2\pi) \, c + \lim_{t \to \infty} \left(\int (1 - f_0^{2t}) \, \psi(\nu) \, d^*\nu - 2c\log t \right) ,$$

where one assumes that $\psi - c \, f_1^{-1}$ is integrable on $\mathbb{R}_+^*$, and

$$f_0(\nu) = \inf(\nu^{1/2}, \nu^{-1/2}) \qquad \forall \nu \in \mathbb{R}_+^* , \qquad f_1 = f_0^{-1} - f_0 \ .$$

The Weil explicit formula is then,

Theorem 1. ([W3]) *With the above notations one has* $S(\mathcal{X}, F) = \Delta(F(|w|) \mathcal{X}(w))$.

Let us make the following change of variables,

$$(12) \qquad\qquad |g|^{-1/2} \, h(g^{-1}) = F(|g|) \, \mathcal{X}_0(g) \,,$$

and rewrite the above equality in terms of h.

By (3) one has,

$$(13) \qquad\qquad \Phi\left(\frac{1}{2} + is\right) = \int_{C_k} F(|g|) |g|^{-is} d^*g \,,$$

thus, in terms of h,

$$(14) \qquad \int h(g) \, \mathcal{X}_1(g) \, |g|^{1/2+is} \, d^*g = \int F(|g^{-1}|) \, \mathcal{X}_0(g^{-1}) \, \mathcal{X}_1(g) \, |g|^{is} \, d^*g \,,$$

which is equal to 0 if $\mathcal{X}_1/C_{k,1} \neq \mathcal{X}_0/C_{k,1}$ and for $\mathcal{X}_1 = \mathcal{X}_0$,

$$(15) \qquad \int h(g)\,\mathcal{X}_0(g)\,|g|^{1/2+is}\,d^*g = \Phi\left(\frac{1}{2}+is\right).$$

We define the Fourier transform on C_k by,

$$(16) \qquad \widehat{h}(\chi,z) = \int_{C^k} h(u)\,\chi(u)\,|u|^z\,d^*u\,.$$

Thus,

$$(17) \qquad \operatorname{Supp}\widehat{h} \subset \mathcal{X}_0 \times \mathbb{R}\,,\ \widehat{h}(\mathcal{X}_0,\rho) = \Phi(\rho)\,,$$

and

$$(18) \qquad S(\mathcal{X}_0,F) = \sum_{\substack{L(\mathcal{X},\rho)=0,\mathcal{X}\in\widehat{C}_{k,1}\\ 0<\operatorname{Re}\rho<1}} \widehat{h}(\mathcal{X},\rho)$$

using a fixed decomposition $C_k = C_{k,1} \times N$.

Let us now evaluate each term in (5).

The first gives $(\log|d^{-1}|)\,h(1)$. One has, using (6) and (12),

$$\langle D, F(|g|)\,\mathcal{X}_0(g)\rangle = \int_{C_k} |g|^{-1/2}\,h(g^{-1})\,(|g|^{1/2}+|g|^{-1/2})\,d^*g$$

$$= \int_{C_k} h(u)\,(1+|u|)\,d^*u = \widehat{h}(0) + \widehat{h}(1)\,,$$

where for the trivial character of $C_{k,1}$ one uses the notation

$$\widehat{h}(z) = \widehat{h}(1,z) \qquad \forall\,z \in \mathbb{C}\,.$$

Thus the first two terms of (5) give

$$(19) \qquad (\log|d^{-1}|)\,h(1) + \widehat{h}(0) + \widehat{h}(1)\,.$$

Let then v be a place of k, one has by (8) and (12),

$$\langle D_v, F(|g|)\,\mathcal{X}_0(g)\rangle = Pfw\int_{k_v^*} \frac{h(u^{-1})}{|1-u|}\,d^*u\,.$$

We can thus write the contribution of the last term of (5) as,

$$(20) \qquad -\sum_v Pfw\int_{k_v^*} \frac{h(u^{-1})}{|1-u|}\,d^*u\,.$$

Thus the equality of Weil can be rewritten as,

$$(21) \qquad \widehat{h}(0) + \widehat{h}(1) - \sum_{\substack{L(\mathcal{X},\rho)=0,\mathcal{X}\in\widehat{C}_{k,1}\\ 0<\operatorname{Re}\rho<1}} \widehat{h}(\mathcal{X},\rho) = (\log|d|)\,h(1)+$$

$$\sum_v Pfw \int_{k_v^*} \frac{h(u^{-1})}{|1-u|} \, d^*u \, .$$

One can slightly improve on this formula and write it in the form,

$$(22) \qquad \sum_{L(\chi,\rho)=0} \widehat{h}(\chi,\rho) - \widehat{h}(0) - \widehat{h}(1) = -\sum_v \int_{k_v^*}' \frac{h(u^{-1})}{|1-u|} \, d^*u \, ,$$

where the principal values $\int'$ depend upon the global Fourier transform. This point was noticed by S. Haran ([H]) and it is crucial for us that these principal values actually coincide ([Co]) with those dictated by the explicit form of the trace formula (cf. sections VIII and IX).

Let us use the geometric dictionary (cf. Introduction) when $\mathrm{char}(k) \neq 0$, namely,

Algebraic Geometry	*Function Theory*
Eigenvalues of action of Frobenius on $H_{\mathrm{et}}^*(\overline{\Sigma}, \mathbb{Q}_\ell)$	Zeros and poles of ζ
Poincaré duality in ℓ-adic cohomology	Functional equation
Lefschetz formula for the Frobenius	Explicit formulas
Castelnuovo positivity	Riemann Hypothesis

The geometric origin of the $-$ sign in (22) becomes clear, (22) is the Lefschetz formula,

$$(23) \qquad \# \text{ of fixed points of } \varphi = \mathrm{Trace}\, \varphi/H^0 - \mathrm{Trace}\, \varphi/H^1 + \mathrm{Trace}\, \varphi/H^2$$

in which the space $H_{\mathrm{et}}^1(\overline{\Sigma}, \mathbb{Q}_\ell)$ which provides the spectral realization of the zeros appears with a $-$ sign. This indicates that the spectral realization of zeros of zeta should be of cohomological nature or to be more specific, that the Polya-Hilbert space should appear as the last term of an exact sequence of Hilbert spaces,

$$(24) \qquad\qquad 0 \to \mathcal{H}_0 \xrightarrow{T} \mathcal{H}_1 \to \mathcal{H} \to 0 \, .$$

The example one can keep in mind for (24) is the assembled Euler complex for a Riemann surface, where $\mathcal{H}_0$ is the *codimension 2 subspace* of the space of differential forms of even degree orthogonal to harmonic forms, where $\mathcal{H}_1$ is the space of 1-forms and where $T = d + d^*$ is the sum of the de Rham coboundary with its adjoint d^*.

Since we want to obtain the spectral interpretation not only for zeta functions but for all L-functions with Grössencharakter we do not expect to have only an action of $\mathbb{Z}$ for $\text{char}(k) > 0$ corresponding to the Frobenius, or of the group $\mathbb{R}_+^*$ if $\text{char}(k) = 0$, but to have the equivariance of (24) with respect to a natural action of the Idele class group $C_k = GL_1(A)/k^*$.

Let $X = A/k^*$ be the Adele class space. Our basic idea is to take for $\mathcal{H}_0$ a suitable completion of the codimension 2 subspace of functions on X such that,

$$(25) \qquad\qquad f(0) = 0 , \ \int f \, dx = 0 ,$$

while $\mathcal{H}_1 = L^2(C_k)$ and T is the restriction map E coming from the inclusion $C_k \to X$, multiplied by $|a|^{1/2}$,

$$(26) \qquad\qquad (Ef)(a) = |a|^{1/2} \, f(a) .$$

The action of C_k is then the obvious one, for $\mathcal{H}_0$

$$(27) \qquad\qquad (U(g)f)(x) = f(g^{-1} x) \qquad \forall \, g \in C_k$$

using the action of C_k on X by multiplication,

$$(28) \qquad\qquad (j, a) \to ja \qquad \forall \, j \in C_k , \ a \in X$$

and similarly the regular representation V for $\mathcal{H}_1$.

There is a subtle point however which is that since C_k is abelian and non compact, its regular representation does not contain any finite dimensional subrepresentation so that the Polya-Hilbert space cannot be a subrepresentation (or unitary quotient) of V. There is an easy way out (which we shall improve shortly) which is to replace the regular representation $L^2(C_k)$ by $L_\delta^2(C_k)$ using the polynomial weight $(\log^2 |a|)^{\delta/2}$, i.e. the norm,

$$(29) \qquad\qquad \|\xi\|_\delta^2 = \int_{C_k} |\xi(a)|^2 \, (1 + \log^2 |a|)^{\delta/2} \, d^*a .$$

The left regular representation V of C_k on $L_\delta^2(C_k)$ is

$$(30) \qquad\qquad (V(a) \, \xi) \, (g) = \xi(a^{-1} g) \qquad \forall \, g, a \in C_k .$$

Note that because of the weight $(1 + \log^2 |x|)^{\delta/2}$, this representation is *not* unitary but it satisfies the growth estimate

$$(31) \qquad\qquad \|V(g)\| = 0 \, (\log |g|)^{\delta/2} \quad \text{when} \quad |g| \to \infty$$

Similarly, we shall construct the Hilbert space L_δ^2 of functions on X with growth indexed by $\delta > 1$. Since X is a quotient space we shall first learn in the usual manifold case how to obtain the Hilbert space $L^2(M)$ of square integrable functions on a manifold M by working only on the universal cover $\widetilde{M}$ with the action of $\Gamma = \pi_1(M)$. Every function $f \in C_c^\infty(\widetilde{M})$ gives rise to a function $\widetilde{f}$ on M by

$$(32) \qquad \widetilde{f}(x) = \sum_{\pi(\widetilde{x})=x} f(\widetilde{x})$$

and all $g \in C^\infty(M)$ appear in this way. Moreover, one can write the Hilbert space inner product $\int_M \widetilde{f}_1(x)\,\widetilde{f}_2(x)\,dx$, in terms of f_1 and f_2 alone. Thus $\|\widetilde{f}\|^2 = \int \left|\sum_{\gamma \in \Gamma} f(\gamma x)\right|^2 dx$ where the integral is performed on a fundamental domain for Γ acting on $\widetilde{M}$. This formula defines a prehilbert space norm on $C_c^\infty(\widetilde{M})$ and $L^2(M)$ is just the completion of $C_c^\infty(\widetilde{M})$ for that norm. Note that any function of the form $f - f_\gamma$ has vanishing norm and hence disappears in the process of completion.

We can now define the Hilbert space $L_\delta^2(X)_0$ as the completion of the codimension 2 subspace

$$(33) \qquad \mathcal{S}(A)_0 = \{f \in \mathcal{S}(A) \;;\; f(0) = 0\ ,\ \int f\,dx = 0\}$$

for the norm $\|\ \|_\delta$ given by

$$(34) \qquad \|f\|_\delta^2 = \int \left|\sum_{q \in k^*} f(qx)\right|^2 (1 + \log^2 |x|)^{\delta/2}\, |x|\, d^*x$$

where the integral is performed on A^*/k^* and d^*x is the multiplicative Haar measure on A^*/k^*. Note that $|qx| = |x|$ for any $q \in k^*$.

The key point is that we use the measure $|x|\,d^*x$ instead of the additive Haar measure dx. Of course for a local field K one has $dx = |x|\,d^*x$ but this fails in the above global situation. Instead one has,

$$(35) \qquad dx = \lim_{\varepsilon \to 0} \varepsilon\,|x|^{1+\varepsilon}\,d^*x\,,$$

but the corresponding divergent normalization coefficient plays no role in computations of adjoints or of traces of operators.

One has a natural representation of C_k on $L_\delta^2(X)_0$ given by (27), and the result is independent of the choice of a lift of j in $J_k = \mathrm{GL}_1(A)$ because the functions $f - f_q$ are in the kernel of the norm. The conditions (33) which define $\mathcal{S}(A)_0$ are invariant under the action of C_k and give the following action of C_k on the 2-dimensional supplement of $\mathcal{S}(A)_0 \subset \mathcal{S}(A)$; this supplement is $\mathbb{C} \oplus \mathbb{C}(1)$ where $\mathbb{C}$ is the trivial C_k module (corresponding to $f(0)$) while the Tate twist $\mathbb{C}(1)$ is the module

$$(36) \qquad (j, \lambda) \to |j|\,\lambda$$

coming from the equality

$$(37) \qquad \int f(j^{-1}x)\,dx = |j| \int f(x)\,dx\,.$$

We let E be the linear isometry from $L^2_\delta(X)_0$ into $L^2_\delta(C_k)$ given by the equality,

$$(38) \qquad\qquad E(f)(g) = |g|^{1/2} \sum_{q \in k^*} f(qg) \qquad \forall g \in C_k .$$

By comparing (29) with (34) we see that E is an isometry and the factor $|g|^{1/2}$ is dictated by comparing the measures $|g|\, d^*g$ of (34) with d^*g of (29).

One has $\quad E(U(a)f)(g) \quad = \quad |g|^{1/2} \sum_{k^*} (U(a)f)(qg) \quad =$
$|g|^{1/2} \sum_{k^*} f(a^{-1}qg) = |a|^{1/2} |a^{-1}g|^{1/2} \sum_{k^*} f(qa^{-1}g) = |a|^{1/2} (V(a)E(f))(g).$
Thus,

$$(39) \qquad\qquad E\,U(a) = |a|^{1/2}\, V(a)\, E .$$

This equivariance shows that the range of E in $L^2_\delta(C_k)$ is a closed invariant subspace for the representation V.

The following theorem and its corollary show that the cokernel $\mathcal{H} = L^2_\delta(C_k)/\operatorname{Im}(E)$ of the isometry E plays the role of the Polya-Hilbert space. Since $\operatorname{Im} E$ is invariant under the representation V we let W be the corresponding representation of C_k on $\mathcal{H}$.

Let $\operatorname{char}(k) = 0$ so that $\operatorname{Mod} k = \mathbb{R}^*_+$ and $C_k = K \times \mathbb{R}^*_+$ where K is the compact group $C_{k,1} = \{a \in C_k \,;\ |a| = 1\}$.

Theorem 2.([Co]) *Let $\delta > 1$, $\mathcal{H}$ be the cokernel of T in $L^2_\delta(C)$ and W the quotient representation of C_k. Let χ be a character of K, $\widetilde{\chi} = \chi \times 1$ the corresponding character of C_k. Let $\mathcal{H}_\chi = \{\xi \in \mathcal{H}\,;\ W(g)\xi = \chi(g)\xi \quad \forall g \in K\}$ and $D_\chi = \lim_{\epsilon \to 0} \frac{1}{\epsilon}(W(e^\epsilon) - 1)$. Then D_χ is an unbounded closed operator with discrete spectrum, $\operatorname{Sp} D_\chi \subset i\,\mathbb{R}$ is the set of imaginary parts of zeros of the L function with Grössencharakter $\widetilde{\chi}$ which have real part $1/2$. Moreover the spectral multiplicity of ρ is the largest integer $n < \frac{1+\delta}{2}$ in $\{1,\dots, \text{multiplicity as a zero of } L\}$.*

Theorem 2 has a similar formulation when the characteristic of k is non zero. The following corollary is valid for global fields k of arbitrary characteristic.

Corollary 3. *For any Schwartz function $h \in \mathcal{S}(C_k)$ the operator $W(h) = \int W(g)\,h(g)\,d^*g$ in $\mathcal{H}$ is of trace class, and its trace is given by*

$$Trace\, W(h) = \sum_{\substack{L\left(\widetilde{\chi}.\frac{1}{2}+\rho\right)=0 \\ \rho \in i\,\mathbb{R}/N^\perp}} \widehat{h}(\widetilde{\chi},\rho)$$

where the multiplicity is counted as in Theorem 2 and where the Fourier transform $\widehat{h}$ of h is defined by,

$$\widehat{h}(\widetilde{\chi},\rho) = \int_{C_k} h(u)\, \widetilde{\chi}(u)\, |u|^\rho\, d^*u .$$

This result is only preliminary because of the unwanted parameter δ which artificially restricts the multiplicities. Let us pursue a little further the analogy between the exact sequence,

$$0 \to L^2_\delta(X)_0 \to L^2_\delta(C_k) \to \mathcal{H} \to 0, \tag{40}$$

and the exact sequence,

$$0 \to \mathcal{H}_0 \overset{T}{\to} \mathcal{H}_1 \to \mathcal{H} \to 0, \tag{41}$$

coming from the assembled Euler complex for a Riemann surface. Thus, here $\mathcal{H}_0$ is the *codimension 2 subspace* of the space of differential forms of even degree orthogonal to harmonic forms, and $\mathcal{H}_1$ is the space of 1-forms while $T = d + d^*$ is the sum of the de Rham coboundary with its adjoint d^*.

In this case Poincaré duality is given by the Hodge $*$ operation, which when multiplied by suitable powers of i satisfies $\tau^2 = 1$ and anticommutes with $T = d + d^*$.

In our case, the analogue of the Hodge $*$ operation is given on $\mathcal{H}_0 = L^2_\delta(X)_0$ by the Fourier transform,

$$(Ff)(x) = \int_A f(y)\,\alpha(xy)\,dy \qquad \forall f \in \mathcal{S}(A)_0. \tag{42}$$

Here, we identified the Abelian group A of Adeles of k with its Pontrjagin dual by means of the pairing $\langle a, b \rangle = \alpha(ab)$, where $\alpha : A \to U(1)$ is a nontrivial character which vanishes on $k \subset A$. Note that such a character is *not canonical*, but that any two such characters α and α' are related by k^*,

$$\alpha'(a) = \alpha(qa) \quad \forall a \in A. \tag{43}$$

It follows that the corresponding Fourier transformations on A are related by

$$\hat{f}' = \hat{f}_q. \tag{44}$$

which, because we take the quotient by k^*, is independent of the choice of additive character α of A. Note also that $F^2 = 1$ on the quotient.

On $\mathcal{H}_1 = L^2_\delta(C_k)$ the Hodge $*$ is given by,

$$(\tau\,\xi)(a) = -\xi(a^{-1}) \qquad \forall a \in C_k. \tag{45}$$

The Poisson formula means exactly that E anticommutes with the $*$ operation.

If we modify the choice of non canonical isomorphism $C_k = K \times \mathbb{R}^*_+$ where K is the compact group $C_{k,1} = \{a \in C_k\,;\, |a| = 1\}$, this modifies the operator D by

$$D' = D - i\,s \tag{46}$$

where $s \in \mathbb{R}$ is determined by the equality

(47) $$\widetilde{\chi}'(g) = \widetilde{\chi}(g)\, |g|^{i\,s} \qquad \forall\, g \in C_k\,.$$

The coherence of the statement of the theorem is insured by the equality

(48) $$L(\widetilde{\chi}', z) = L(\widetilde{\chi}, z + i\,s) \qquad \forall\, z \in \mathbb{C}.$$

When the zeros of L have multiplicity and δ is large enough the operator D is *not* semisimple and has a non trivial Jordan form.

The proof of theorem 2 ([Co]) is based on the distribution theoretic interpretation by A. Weil [W2] of the idea of Tate and Iwasawa on the functional equation. Our construction should be compared with [Bg] and [Z].

Since we obtain the Hilbert space $L^2_\delta(X)_0$ by imposing two linear conditions on $\mathcal{S}(A)$,

(49) $$0 \to \mathcal{S}(A)_0 \to \mathcal{S}(A) \xrightarrow{L} \mathbb{C} \oplus \mathbb{C}(1) \to 0$$

we shall define $L^2_\delta(X)$ so that it fits in an exact sequence of C_k-modules

(50) $$0 \to L^2_\delta(X)_0 \to L^2_\delta(X) \to \mathbb{C} \oplus \mathbb{C}(1) \to 0\,.$$

(Note that, as in the case of Riemann surfaces, the $*$ operation (42) does exchange the two modules $\mathbb{C}$ and $\mathbb{C}(1)$). We can then use the exact sequence of C_k-modules

(51) $$0 \to L^2_\delta(X)_0 \to L^2_\delta(C_k) \to \mathcal{H} \to 0$$

together with Corollary 3 to compute in a formal manner what the character of the module $L^2_\delta(X)$ should be. We obtain,

(52) $$\text{``Trace''} \ (U(h)) = \widehat{h}(0) + \widehat{h}(1) - \sum_{\substack{L(\chi,\rho)=0 \\ \mathrm{Re}\,\rho = \frac{1}{2}}} \widehat{h}(\chi,\rho) + \infty\, h(1)$$

where $\widehat{h}(\chi,\rho)$ is defined by Corollary 3 and

(53) $$U(h) = \int_{C_k} U(g)\, h(g)\, d^*g$$

while the test function h is in a suitable function space. Note that the trace on the left hand side of (52) only makes sense after a suitable regularisation since the left regular representation of C_k is not traçable. This situation is similar to the one encountered by Atiyah and Bott ([AB]) in their proof of the Lefschetz formula. We shall first learn how to compute in a formal manner the above trace from the fixed points of the action of C_k on X. In sections VIII and IX, we shall show how to regularize the trace and completely eliminate the parameter δ.

VI The distribution trace formula for flows on manifolds

In order to understand how the left hand side of V(52) should be computed we shall first give an account of the proof of the usual Lefschetz formula by Atiyah-Bott ([AB]) and describe the computation of the distribution theoretic trace for flows on manifolds, which is a variation on the theme of [AB] and is due to Guillemin-Sternberg [GS]. We first recall for the convenience of the reader the coordinate free treatment of distributions of [GS].

Given a vector space E over $\mathbb{R}$, $\dim E = n$, a density is a map, $\rho \in |E^*|$,

$$(1) \qquad \rho : \wedge^n E \to \mathbb{C}$$

such that $\rho(\lambda v) = |\lambda|\, \rho(v) \qquad \forall \lambda \in \mathbb{R}, \qquad \forall v \in \wedge^n E.$

Given a linear map $T : E \to F$ we let $|T^*| : |F^*| \to |E^*|$ be the corresponding linear map, it depends contravariantly on T.

A smooth compactly supported density $\rho \in C_c^\infty(M, |T^*M|)$ on an arbitrary manifold M has a canonical integral,

$$(2) \qquad \int \rho \in \mathbb{C}.$$

One defines the generalized sections of a vector bundle L on M as the dual space of $C_c^\infty(M, L^* \otimes |T^*M|)$

$$(3) \qquad C^{-\infty}(M, L) = \text{ dual of } C_c^\infty(M, L^* \otimes |T^*M|)$$

where L^* is the dual bundle. One has a natural inclusion,

$$(4) \qquad C^\infty(M, L) \subset C^{-\infty}(M, L)$$

given by the pairing

$$(5) \qquad \sigma \in C^\infty(M, L) ,\ s \in C_c^\infty(M, L^* \otimes |T^*M|) \to \int \langle s, \sigma \rangle$$

where $\langle s, \sigma \rangle$ is viewed as a density, $\langle s, \sigma \rangle \in C_c^\infty(M, |T^*M|)$.

One has a similar notion of generalized section with compact support.

Given a smooth map $\varphi : X \to Y$, then if φ is *proper*, it gives a (contravariantly) associated map

$$(6) \qquad \varphi^* : C_c^\infty(Y, L) \to C_c^\infty(X, \varphi^*(L)) ,\ (\varphi^* \xi)(x) = \xi(\varphi(x))$$

where $\varphi^*(L)$ is the pull back of the vector bundle L.

Thus, given a linear form on $C_c^\infty(X, \varphi^*(L))$ one has a (covariantly) associated linear form on $C_c^\infty(Y, L)$. In particular with L trivial we see that generalized densities $\rho \in C^{-\infty}(X, |T^*X|)$ pushforward,

$$(7) \qquad\qquad \varphi_*(\rho) \in C^{-\infty}(Y, |T^*Y|)$$

with $\langle \varphi_*(\rho), \xi \rangle = \langle \rho, \varphi^* \xi \rangle \qquad \forall \xi \in C_c^\infty(X)$.

This gives the natural functoriality of generalized sections, they pushforward under proper maps. However under suitable transversality conditions which are automatic for submersions, generalized sections also pull back. For instance, if φ is a fibration and $\rho \in C_c^\infty(X, |T^*X|)$ is a density then one can integrate ρ along the fibers, the obtained density on Y, $\varphi_*(\rho)$ is given as in (7) by

$$(8) \qquad\qquad \langle \varphi_*(\rho), f \rangle = \langle \rho, \varphi^* f \rangle \qquad \forall f \in C^\infty(Y).$$

The point is that the result is not only a generalized section but a smooth section $\varphi_*(\rho) \in C_c^\infty(Y, |T^*Y|)$.

It follows that if $f \in C^{-\infty}(Y)$ is a generalized function, then one obtains a generalized function $\varphi^*(f)$ on X by,

$$(9) \qquad\qquad \langle \varphi^*(f), \rho \rangle = \langle f, \varphi_*(\rho) \rangle \qquad \forall \rho \in C_c^\infty(X, |T^*X|).$$

In general, the pullback $\varphi^*(f)$ of a generalized function f, continues to make sense provided the following transversality condition holds,

$$(10) \qquad\qquad d(\varphi^*(l)) \neq 0 \qquad \forall l \in WF(f).$$

where $WF(f)$ is the wave front set of f ([GS]).

Next, let us recall the construction ([GS]) of the generalized section of a vector bundle L on a manifold X associated to a submanifold $Z \subset X$ and a symbol,

$$(11) \qquad\qquad \sigma \in C^\infty(Z, L \otimes |N_Z|).$$

where N_Z is the normal bundle of Z. The construction is the same as that of the current of integration on a cycle. Given $\xi \in C_c^\infty(X, L^* \otimes |T^*X|)$, the product $\sigma \xi / Z$ is a density on Z, since it is a section of $|T_Z^*| = |T_X^*| \otimes |N_Z|$. One can thus integrate it over Z.

When $Z = X$ one has $N_Z = \{0\}$ and $|N_Z|$ has a canonical section, so that the current associated to σ is just given by (5).

Now let $\varphi : X \to Y$ with Z a submanifold of Y and σ as in (11).

Let us assume that φ is transverse to Z, so that for each $x \in X$ with $y = \varphi(x) \in Z$ one has

$$(12) \qquad\qquad \varphi_*(T_x) + T_{\varphi(x)}(Z) = T_y Y.$$

Let

(13) $$\tau_x = \{X \in T_x \ , \ \varphi_*(X) \in T_y(Z)\} \, .$$

Then φ_* gives a canonical isomorphism,

(14) $$\varphi_* : T_x(X)/\tau_x \simeq T_y(Y)/T_y(Z) = N_y(Z) \, .$$

And $\varphi^{-1}(Z)$ is a submanifold of X of the same codimension as Z with a natural isomorphism of normal bundles

(15) $$\varphi_* : N_{\varphi^{-1}(Z)} \simeq \varphi^* \, N_Z \, .$$

In particular, given a (generalized) δ-section of a bundle L with support Z and symbol $\sigma \in C^\infty(Z, L \otimes |N_Z|)$ one has a corresponding symbol on $\varphi^{-1}(Z)$ given by

(16) $$\varphi^* \, \sigma(x) = |(\varphi_*)^{-1}|\sigma(\varphi(x)) \in (\varphi^* \, L)_x \otimes |N_x|$$

using the inverse of the isomorphism (15), which requires the transversality condition.

Now for any δ-section associated to Z, σ, the wave front set is contained in the conormal bundle of the submanifold Z which shows that if φ is transverse to Z the pull back $\varphi^* \, \delta_{Z,\sigma}$ of the distribution on Y associated to Z, σ makes sense, it is equal to $\delta_{\varphi^{-1}(Z),\varphi^*(\sigma)}$.

Let us now recall the formulation ([GS]) of the Schwartz kernel theorem. One considers a continuous linear map,

(17) $$T : C_c^\infty(Y) \to C^{-\infty}(X) \, ,$$

the statement is that one can write it as

(18) $$(T \, \xi)\,(x) = \int k(x,y)\, \xi(y)\, dy$$

where $k(x,y)\, dy$ is a generalized section,

(19) $$k \in C^{-\infty}(X \times Y \ , \ \mathrm{pr}_Y^*(|T^*Y|)) \, .$$

Let $f : X \to Y$ be a smooth map, and $T = f^*$ the operator

(20) $$(T \, \xi)\,(x) = \xi\,(f(x)) \qquad \forall \xi \in C_c^\infty(Y) \, .$$

The corresponding k is the δ-section associated to the submanifold of $X \times Y$ given by

(21) $$\mathrm{Graph}(f) = \{(x, f(x)) \ ; \ x \in X\} = Z$$

and its symbol, $\sigma \in C^\infty(Z, \mathrm{pr}_Y^*(|T^*Y|) \otimes |N_Z|)$ is obtained as follows.

Given $\xi \in T_x^*(X)$, $\eta \in T_y^*(Y)$ one has $(\xi, \eta) \in N_Z^*$ iff it is orthogonal to $(v, f_* v)$ for any $v \in T_x(X)$, i.e. $\langle v, \xi \rangle + \langle f_* v, \eta \rangle = 0$ so that

$$\text{(22)} \qquad \xi = -f_*^t \eta \, .$$

Thus one has a canonical isomorphism $j : T_y^*(Y) \simeq N_Z^*$, $\eta \overset{j}{\to} (-f_*^t \eta, \eta)$. The transposed $(j^{-1})^t$ is given by $(j^{-1})^t(Y) = $ class of $(0, Y)$ in N_Z, $\forall Y \in T_y(Y)$. One has,

$$\text{(23)} \qquad \sigma = |j^{-1}| \in C^\infty(Z, \mathrm{pr}_Y^*(|T^*Y|) \otimes |N_Z|) \, .$$

We denote the corresponding δ-distribution by

$$\text{(24)} \qquad k(x, y) \, dy = \delta(y - f(x)) \, dy \, .$$

One then checks the formula,

$$\text{(25)} \qquad \int \delta(y - f(x)) \, \xi(y) \, dy = \xi(f(x)) \qquad \forall \xi \in C_c^\infty(Y) \, .$$

Let us now consider a manifold M with a flow F_t

$$\text{(26)} \qquad F_t(x) = \exp(t v) \, x \qquad v \in C^\infty(M, T_M)$$

and the corresponding map f,

$$\text{(27)} \qquad f : M \times \mathbb{R} \to M \, , \ f(x, t) = F_t(x) \, .$$

We apply the above discussion with $X = M \times \mathbb{R}$, $Y = M$. The graph of f is the submanifold Z of $X \times Y$,

$$\text{(28)} \qquad Z = \{(x, t, y) \; ; \; y = F_t(x)\} \, .$$

One lets φ be the diagonal map,

$$\text{(29)} \qquad \varphi(x, t) = (x, t, x) \, , \ \varphi : M \times \mathbb{R} \to X \times Y$$

and the first issue is the transversality $\varphi \pitchfork Z$.

We thus need to consider (12) for each (x, t) such that $\varphi(x, t) \in Z$, i.e. such that $x = F_t(x)$. One looks at the image by φ_* of the tangent space $T_x M \times \mathbb{R}$ to $M \times \mathbb{R}$ at (x, t). One lets ∂_t be the natural vector field on $\mathbb{R}$. The image of $(X, \lambda \partial_t)$ is $(X, \lambda \partial_t, X)$ for $X \in T_x M, \lambda \in \mathbb{R}$. Dividing the tangent space of $M \times \mathbb{R} \times M$ by the image of φ_* one gets an isomorphism,

$$\text{(30)} \qquad (X, \lambda \partial_t, Y) \to Y - X$$

with $T_x M$. The tangent space to Z is $\{(X', \mu \, \partial_t, (F_t)_* \, X' + \mu \, v_{F_t(x)}); \ X' \in T_x M, \mu \in \mathbb{R}\}$. Thus the transversality condition means that every element of $T_x M$ is of the form

(31) $(F_t)_* \, X - X + \mu \, v_x \qquad X \in T_x M \ , \ \mu \in \mathbb{R}.$

One has

(32) $(F_t)_* \, \mu \, v_x = \mu \, v_x$

so that $(F_t)_*$ defines a quotient map, the Poincaré return map

(33) $P : T_x / \mathbb{R} v_x \to T_x / \mathbb{R} v_x = N_x$

and the transversality condition (31) means exactly,

(34) $1 - P \qquad$ is invertible.

Let us make this hypothesis and compute the symbol σ of the distribution,

(35) $\tau = \varphi^* (\delta(y - F_t(x)) \, dy) \, .$

First, as above, let $W = \varphi^{-1}(Z) = \{(x,t) \ ; \ F_t(x) = x\}$. The codimension of $\varphi^{-1}(Z)$ in $M \times \mathbb{R}$ is the same as the codimension of Z in $M \times \mathbb{R} \times M$ so it is $\dim M$ which shows that $\varphi^{-1}(Z)$ is 1-dimensional. If $(x,t) \in \varphi^{-1}(Z)$ then $(F_s(x), t) \in \varphi^{-1}(Z)$. Thus, if we assume that v does not vanish at x, the map,

(36) $(x,t) \overset{q}{\to} t$

is locally constant on the connected component of $\varphi^{-1}(Z)$ containing (x,t).
This allows to identify the transverse space to $W = \varphi^{-1}(Z)$ as the product,

(37) $N_{x,t}^W \simeq N_x \times \mathbb{R}$

where to $(X, \lambda \, \partial_t) \in T_{x,t}(M \times \mathbb{R})$ we associate the pair $(\widetilde{X}, \lambda)$ given by the class of X in $N_x = T_x / \mathbb{R} v_x$ and $\lambda \in \mathbb{R}$.

The symbol σ of the distribution (35) is a smooth section of $|N^W|$ tensored by the pull back $\varphi^*(L)$ where $L = \mathrm{pr}_Y^* \, |T_M^*|$, and one has

(38) $\varphi^*(L) \simeq |p^* \, T_M^*|$

where

(39) $p(x,t) = x \qquad \forall \, (x,t) \in M \times \mathbb{R} \, .$

To compute σ one needs the isomorphism,

(40) $N_{(x,t)}^W \overset{\varphi}{\to} T_{\varphi(x,t)}(M \times \mathbb{R} \times M) / T_{\varphi(x,t)}(Z) = N^Z \, .$

The map $\varphi_* : N^W_{x,t} \to N^Z$ is given by

$$(41) \qquad \varphi_*(X, \lambda\, \partial_t) = (1 - (F_t)_*)\, X - \lambda\, v \qquad X \in N_x\ ,\ \lambda \in \mathbb{R}$$

and the symbol σ is just

$$(42) \qquad \sigma = |\varphi_*^{-1}| \in |p^*\, T^*_M| \otimes |N^W|\, .$$

Let us now consider the second projection,

$$(43) \qquad q(x,t) = t \in \mathbb{R}$$

and compute the pushforward $q_*(\tau)$ of the distribution τ.

By construction $q_*(\tau)$ is a generalized function.

We first look at the contribution of a periodic orbit, the corresponding part of $\varphi^{-1}(Z)$ is of the form,

$$(44) \qquad \varphi^{-1}(Z) = V \times \Gamma \subset M \times \mathbb{R}$$

where Γ is a discrete cocompact subgroup of $\mathbb{R}$, while $V \subset M$ is a one dimensional compact submanifold of M.

To compute $q_*(\tau)$, we let $h(t)\, |dt|$ be a 1-density on $\mathbb{R}$ and pull it back by q as the section on $M \times \mathbb{R}$ of the bundle $q^*\, |T^*|$,

$$(45) \qquad \xi(x,t) = h(t)\, |dt|\, .$$

We now need to compute $\int_{\varphi^{-1}(Z)} \xi\, \sigma$. We can look at the contribution of each component: $V \times \{T\}$, $T \in \Gamma$.

One gets ([GS]),

$$(46) \qquad T^\# \frac{1}{|1 - P_T|}\, h(T)\, ,$$

where $T^\#$ is the length of the primitive orbit or equivalently the covolume of Γ in $\mathbb{R}$ for the Haar measure $|dt|$. We can thus write the contributions of the periodic orbits as

$$(47) \qquad \sum_{\gamma_p} \sum_{\Gamma} \mathrm{Covol}(\Gamma)\, \frac{1}{|1 - P_T|}\, h(T)\, ,$$

where the test function h vanishes at 0.

The next case to consider is when the vector field v_x has an isolated 0, $v_{x_0} = 0$. In that case, the transversality condition (31) becomes

$$(48) \qquad 1 - (F_t)_* \quad \text{invertible (at } x_0)\, .$$

One has $F_t(x_0) = x_0$ for all $t \in \mathbb{R}$ and now the relevant component of $\varphi^{-1}(Z)$ is $\{x_0\} \times \mathbb{R}$. The transverse space N^W is identified with T_x and the map $\varphi_* : N^W \simeq N^Z$ is given by:

$$(49) \qquad\qquad \varphi_* = 1 - (F_t)_* \,.$$

Thus the symbol σ is the scalar function $|1 - (F_t)_*|^{-1}$. The generalized section $q_* \, \varphi^*(\delta(y - F_t(x))\, dy)$ is the function, $t \to |1 - (F_t)_*|^{-1}$. We can thus write the contribution of the zeros of the flow as ([GS]),

$$(50) \qquad\qquad \sum_{zeros} \int \frac{h(t)}{|1 - (F_t)_*|} \, dt$$

where h is a test function vanishing at 0.

We can thus collect the contributions 47 and 50 as

$$(51) \qquad\qquad \sum_{\gamma} \int_{I_\gamma} \frac{h(u)}{|1 - (F_u)_*|} \, d^*u$$

where h is as above, I_γ is the isotropy group of the periodic orbit γ, the haar measure d^*u on I_γ is normalised so that the covolume of I_γ is equal to one and we still write $(F_u)_*$ for its restriction to the transverse space of γ.

VII The global case, and the formal trace computation.

We shall consider the action of C_k on X and write down the analogue of VI (51) for the distribution trace formula.

Both X and C_k are defined as quotients and we let

$$(1) \qquad \pi : A \to X \,, \ c : \mathrm{GL}_1(A) \to C_k$$

be the corresponding quotient maps.

As above we consider the graph Z of the action

$$(2) \qquad f : X \times C_k \to X \,, \ f(x, \lambda) = \lambda x$$

and the diagonal map

$$(3) \qquad \varphi : X \times C_k \to X \times C_k \times X \qquad \varphi(x, \lambda) = (x, \lambda, x) \,.$$

We first investigate the fixed points, $\varphi^{-1}(Z)$, i.e. the pairs $(x, \lambda) \in X \times C_k$ such that $\lambda x = x$. Let $x = \pi(\tilde{x})$ and $\lambda = c(j)$. Then the equality $\lambda x = x$ means that $\pi(j\tilde{x}) = \pi(\tilde{x})$ thus there exists $q \in k^*$ such that with $\tilde{j} = qj$, one has

$$(4) \qquad \tilde{j}\tilde{x} = \tilde{x} \,.$$

Recall now that A is the restricted direct product $A = \prod_{res} k_v$ of the local fields k_v obtained by completion of k with respect to the place v. The equality (4) means that $\tilde{j}_v \tilde{x}_v = \tilde{x}_v$, thus, if $\tilde{x}_v \neq 0$ for all v it follows that $\tilde{j}_v = 1 \ \forall v$ and $\tilde{j} = 1$. This shows that the projection of $\varphi^{-1}(Z) \cap C_k \backslash \{1\}$ on X is the union of the hyperplanes

$$(5) \qquad \cup H_v \,; \ H_v = \pi(\tilde{H}_v) \,, \ \tilde{H}_v = \{x \,; \ x_v = 0\} \,.$$

Each $\tilde{H}_v$ is closed in A and is invariant under multiplication by elements of k^*. Thus each H_v is a closed subset of X and one checks that it is the closure of the orbit under C_k of any of its generic points

$$(6) \qquad x \,, \ x_u = 0 \quad \Longleftrightarrow \quad u = v \,.$$

For any such point x, the isotropy group I_x is the image in C_k of the multiplicative group k_v^*,

$$(7) \qquad I_x = k_v^*$$

by the map $\lambda \in k_v^* \to (1, \ldots, 1, \lambda, 1, \ldots)$. This map already occurs in class field theory (cf [W1]) to relate the local theory to the global one.

Both groups k_v^* and C_k are commensurable to $\mathbb{R}_+^*$ by the module homomorphism, which is proper with cocompact range,

$$(8) \qquad G \xrightarrow{|\,|} \mathbb{R}_+^* .$$

Since the restriction to k_v^* of the module of C_k is the module of k_v^*, it follows that

$$(9) \qquad I_x \text{ is a cocompact subgroup of } C_k .$$

This allows to normalize the respective Haar measures in such a way that the covolume of I_x is 1. This is in fact insured by the canonical normalization of the Haar measures of modulated groups ([W3]),

$$(10) \qquad \int_{|g| \in [1, \Lambda]} d^* g \sim \log \Lambda \quad \text{when} \quad \Lambda \to +\infty .$$

It is important to note that though I_x is cocompact in C_k, the orbit of x is not closed and one needs to close it, the result being H_v. We shall learn how to justify this point later in section VIII, in the similar situation of the action of C_S on X_S. We can now in view of the results of the two preceding sections, write down the contribution of each H_v to the distributional trace;

Since $\tilde{H}_v$ is a hyperplane, we can identify the transverse space N_x to H_v at x with the quotient

$$(11) \qquad N_x = A/\tilde{H}_v = k_v,$$

namely the additive group of the local field k_v. Given $j \in I_x$ one has $j_u = 1 \ \forall u \neq v$, and $j_v = \lambda \in k_v^*$. The action of j on A is linear and fixes x, thus the action on the transverse space N_x is given by

$$(12) \qquad (\lambda, a) \to \lambda a \quad \forall a \in k_v.$$

We can thus proceed with some faith and write down the contribution of H_v to the distributional trace in the form,

$$(13) \qquad \int_{k_v^*} \frac{h(\lambda)}{|1 - \lambda|} d^* \lambda,$$

where h is a test function on C_k which vanishes at 1. We now have to take care of a discrepancy in notation with the fifth section, where we used the symbol $U(j)$ for the operation

$$(14) \qquad (U(j)f)(x) = f(j^{-1}x)$$

whereas we use j in the above discussion. This amounts to replace the test function $h(u)$ by $h(u^{-1})$ and we thus obtain as a formal analogue of VI(51) the following expression for the distributional trace

$$(15) \qquad \text{"Trace" } (U(h)) = \sum_v \int_{k_v^*} \frac{h(u^{-1})}{|1-u|} \, d^* u \, .$$

Now the right-hand side of (15) is, when restricted to the hyperplane $h(1) = 0$, the distribution obtained by André Weil ([W3]) (cf. Theorem V.1) as the synthesis of the explicit formulas of number theory for all L-functions with Grössencharakter. In particular we can rewrite it as

$$(16) \qquad \hat{h}(0) + \hat{h}(1) - \sum_{L(\chi,\rho)=0} \hat{h}(\chi,\rho) + \infty \, h(1)$$

where this time the restriction $\text{Re}(\rho) = \frac{1}{2}$ has been eliminated.

Thus, equating (52) of section V and (16) for $h(1) = 0$ would yield the desired information on the zeros. Of course, this does require first eliminating the role of δ, and (as in [AB]) to prove that the distributional trace coincides with the ordinary operator theoretic trace on the cokernel of E. This is achieved for the usual set-up of the Lefschetz fixed point theorem by the use of families.

A very important property of the right hand side of (15) is the following "Hadamard positivity": If the test function $h, h(1) = 0$ is positive,

$$(17) \qquad h(u) \geq 0 \quad \forall \, u \in C_k$$

then the right-hand side is *positive*. This indicated from the very start that in order to obtain the Polya-Hilbert space from the Riemann flow, it is *not* quantization that should be involved but simply the passage to the L^2 space, $X \to L^2(X)$. Indeed the positivity of (17) is typical of *permutation matrices* rather than of quantization. This distinction plays a crucial role in the above discussion of the trace formula, in particular the expected trace formula is not a semi-classical formula but a Lefschetz formula in the spirit of [AB].

The above discussion is *not* a rigorous justification of this formula. The first obvious obstacle is that the distributional trace is only formal and to give it a rigorous meaning tied up to Hilbert space operators, one needs as we shall see in section VIII, to perform a cutoff. The second difficulty comes from the presence of the parameter δ as a label for the Hilbert space, while δ does not appear in the trace formula. As we shall see in the next two sections the cutoff will completely eliminate the role of δ, and we shall nevertheless show (by proving positivity of the Weil distribution) that the validity of the (δ independent) trace formula is equivalent to the Riemann Hypothesis for all Grössencharakters of k.

VIII The trace formula in the S-local case.

In the formal trace computation of section VII, we skiped over the difficulties inherent to the tricky structure of the space X. In order to understand how to handle trace formulas on such spaces we shall consider the slightly simpler situation which arises when one only considers a finite set S of places of k. As soon as the cardinality of S is larger than 3, the corresponding space X_S does share most of the tricky features of the space X. In particular it is no longer of type I in the sense of Noncommutative Geometry.

We shall nevertheless describe a precise general result ([Co] theorem 4) which shows that the above handling of periodic orbits and of their contribution to the trace is the correct one. It will in particular show why the orbit of the fixed point 0, or of elements $x \in A$, such that x_v vanishes for at least two places do not contribute to the trace formula. At the same time, we shall handle the lack of transversality when $h(1) \neq 0$.

Let us begin by the local case. let K be a local field. We deal directly with the following operator in $L^2(K)$,

$$(1) \qquad U(h) = \int h(\lambda) \, U(\lambda) \, d^*\lambda \,,$$

where the scaling operator $U(\lambda)$ is defined by

$$(2) \qquad (U(\lambda)\,\xi)(x) = \xi(\lambda^{-1}\,x) \qquad \forall\, x \in K$$

and where the multiplicative Haar measure $d^*\lambda$ is normalized by,

$$(3) \qquad \int_{|\lambda|\in[1,\Lambda]} d^*\lambda \sim \log\Lambda \qquad \text{when } \Lambda \to \infty \,.$$

To understand the "trace" of $U(h)$ we shall proceed as in the Selberg trace formula ([Se]) and use a cutoff. In physics terminology the divergence of the trace is both infrared and ultraviolet. To perform an infrared cutoff, we use the orthogonal projection P_Λ onto the subspace,

$$(4) \qquad P_\Lambda = \{\xi \in L^2(K)\,;\; \xi(x) = 0 \qquad \forall x\,, \; |x| > \Lambda\}\,.$$

Thus, P_Λ is the multiplication operator by the function ρ_Λ, where $\rho_\Lambda(x) = 1$ if $|x| \leq \Lambda$, and $\rho(x) = 0$ for $|x| > \Lambda$. This gives an infrared cutoff and to get an ultraviolet cutoff we use $\widehat{P}_\Lambda = FP_\Lambda F^{-1}$ where F is the Fourier transform (which depends upon the basic character α). We let

(5) $R_\Lambda = \widehat{P}_\Lambda\, P_\Lambda\,.$

One then obtains ([Co]),

Theorem 1. *Let K be a local field with basic character α. Let $h \in \mathcal{S}(K^*)$ have compact support. Then $R_\Lambda\, U(h)$ is a trace class operator and when $\Lambda \to \infty$, one has*

$$\mathrm{Trace}\,(R_\Lambda\, U(h)) = 2h(1)\log' \Lambda + \int' \frac{h(u^{-1})}{|1-u|}\, d^*u + o(1)$$

where $2\log' \Lambda = \int_{\lambda \in K^*,\, |\lambda|\in[\Lambda^{-1},\Lambda]} d^*\lambda$, and the principal value $\int'$ is uniquely determined by the pairing with the unique distribution on K which agrees with $\frac{du}{|1-u|}$ for $u \neq 1$ and whose Fourier transform vanishes at 1.

As it turns out ([Co]), this principal value agrees with that of Weil (cf. section V) for the choice of F associated to the standard character of K.

Let us now describe the reduced framework for the trace formula. We now let k be a global field and S a finite set of places of k containing all infinite places. The group O_S^* of S-units is defined as the subgroup of k^*,

(6) $O_S^* = \{q \in k^*, |q_v| = 1, v \notin S\}$

It is cocompact in J_S^1 where,

(7) $J_S = \prod_{v \in S} k_v^*$

and,

(8) $J_S^1 = \{j \in J_S, |j| = 1\}.$

Thus the quotient group $C_S = J_S/O_S^*$ plays the same role as C_k, and acts on the quotient X_S of $A_S = \prod_{v \in S} k_v$ by O_S^*.

To keep in mind a simple example, one can take $k = \mathbb{Q}$, while S consists of the three places 2, 3, and ∞. One checks in this example that the topology of X_S is not of type I since for instance the group $O_S^* = \{\pm 2^n 3^m; n, m \in \mathbb{Z}\}$ acts ergodically on $\{0\} \times \mathbb{R} \subset A_S$.

We normalize the multiplicative Haar measure $d^*\lambda$ of C_S by,

(9) $\int_{|\lambda|\in[1,\Lambda]} d^*\lambda \sim \log \Lambda \qquad$ when $\Lambda \to \infty\,,$

and normalize the multiplicative Haar measure $d^*\lambda$ of J_S so that it agrees with the above on a fundamental domain D for the action of O_S^* on J_S.

There is no difficulty in defining the Hilbert space $L^2(X_S)$ of square integrable functions on X_S. We proceed as in section V (without the δ), and complete (and separate) the Schwartz space $\mathcal{S}(A_S)$ for the pre-Hilbert structure given by,

$$\|f\|^2 = \int \Big| \sum_{q\in O_S^*} f(qx)\Big|^2 |x|\, d^*x \tag{10}$$

where the integral is performed on C_S or equivalently on a fundamental domain D for the action of O_S^* on J_S. To show that (10) makes sense, one proves that for $f \in \mathcal{S}(A_S)$, the function $E_0(f)(x) = \sum_{q\in O_S^*} f(qx)$ is bounded above by a power of $Log|x|$ when $|x|$ tends to zero. To see this when f is the characteristic function of $\{x \in A_S, |x_v| \le 1, \forall v \in S\}$, one uses the cocompactness of O_S^* in J_S^1, to replace the sum by an integral. The latter is then comparable to,

$$\int_{u_i \ge 0, \sum u_i = -Log|x|} \prod du_i, \tag{11}$$

where the index i varies in S. The general case follows.

The scaling operator $U(\lambda)$ is defined by,

$$(U(\lambda)\xi)(x) = \xi(\lambda^{-1}x) \qquad \forall x \in A_S \tag{12}$$

and the same formula, with $x \in X_S$ defines its action on $L^2(X_S)$. Given a smooth compactly supported function h on C_S, $U(h) = \int h(g)U(g)d^*g$ makes sense as an operator acting on $L^2(X_S)$.

We shall first see that the Fourier transform F on $\mathcal{S}(A_S)$ does extend to a unitary operator on the Hilbert space $L^2(X_S)$.

Lemma 2. ([Co]) a) *For any $f_i \in \mathcal{S}(A_S)$ the series $\sum_{O_S^*} \langle f_1, U(q) f_2\rangle_A$ of inner products in $L^2(A_S)$ converges geometrically on the abelian finitely generated group O_S^*. Moreover its sum is equal to the inner product of f_1 and f_2 in the Hilbert space $L^2(X_S)$.*

b) Let $\alpha = \prod \alpha_v$ be a basic character of the additive group A_S and F the corresponding Fourier transformation. The map $f \to F(f)$, $f \in \mathcal{S}(A_S)$ extends uniquely to a unitary operator in the Hilbert space $L^2(X_S)$.

Now exactly as above for the case of local fields (theorem 1), we need to use a cutoff. For this we use the orthogonal projection P_Λ onto the subspace,

$$P_\Lambda = \{\xi \in L^2(X_S)\,;\ \xi(x) = 0 \qquad \forall x\,,\ |x| > \Lambda\}. \tag{13}$$

Thus, P_Λ is the multiplication operator by the function ρ_Λ, where $\rho_\Lambda(x) = 1$ if $|x| \le \Lambda$, and $\rho(x) = 0$ for $|x| > \Lambda$. This gives an infrared cutoff and to get an ultraviolet cutoff we use $\widehat{P}_\Lambda = FP_\Lambda F^{-1}$ where F is the Fourier transform (lemma 1) which depends upon the choice of the basic character $\alpha = \prod \alpha_v$. We let

$$R_\Lambda = \widehat{P}_\Lambda P_\Lambda. \tag{14}$$

One then gets ([Co]),

Theorem 3. *Let A_S be as above, with basic character $\alpha = \prod \alpha_v$. Let $h \in \mathcal{S}(C_S)$ have compact support. Then when $\Lambda \to \infty$, one has*

$$\operatorname{Trace}\left(R_\Lambda\, U(h)\right) = 2h(1)\log' \Lambda + \sum_{v \in S} \int_{k_v^*}' \frac{h(u^{-1})}{|1-u|}\, d^*u + o(1)$$

where $2\log' \Lambda = \int_{\lambda \in C_S,\, |\lambda| \in [\Lambda^{-1}, \Lambda]} d^*\lambda$, each k_v^* is embedded in C_S by the map $u \to (1, 1, ..., u, ..., 1)$ and the principal value $\int'$ is uniquely determined by the pairing with the unique distribution on k_v which agrees with $\frac{du}{|1-u|}$ for $u \neq 1$ and whose Fourier transform relative to α_v vanishes at 1.

Let us now discuss the global trace formula.

IX The global trace formula and the geometric dictionary.

The main difficulty created by the parameter δ in Theorem V.2 is that the formal trace computation of section VII is independent of δ, and thus cannot give in general the expected value of the trace of corollary V.3, since in the latter each critical zero ρ is counted with a multiplicity equal to the largest integer $n < \frac{1+\delta}{2}$, $n \leq$ multiplicity of ρ as a zero of L. In particular for L functions with multiple zeros, the δ-dependence of the spectral side is nontrivial. It is also clear that the function space $L^2_\delta(X)$ artificially eliminates the non-critical zeros by the introduction of the δ.

As we shall see, all these problems are eliminated by the cutoff. The latter will be performed directly on the Hilbert space $L^2(X)$ so that the only value of δ that we shall use is $\delta = 0$. All zeros will play a role in the spectral side of the trace formula, but while the critical zeros will appear per-se, the non critical ones will appear as resonances and enter in the trace formula through their harmonic potential with respect to the critical line. Thus the spectral side is now entirely canonical and independent of δ, and by proving positivity of the Weil distribution, we shall show that its equality with the geometric side, i.e. the global analogue of Theorem VIII.3, is equivalent to the Riemann Hypothesis for all L-functions with Grössencharakter.

The Abelian group A of Adeles of k is its own Pontrjagin dual by means of the pairing

$$(1) \qquad\qquad \langle a, b \rangle = \alpha(ab)$$

where $\alpha : A \to U(1)$ is a nontrivial character which vanishes on $k \subset A$.

We fix the additive character α as above, $\alpha = \prod \alpha_v$ and let d be a differential idele,

$$(2) \qquad\qquad \alpha(x) = \alpha_0(d\,x) \quad \forall x \in A,$$

where $\alpha_0 = \prod \alpha_{0,v}$ is the product of the local normalized additive characters (cf [W1]). We let S_0 be the finite set of places where α_v is ramified.

We shall first concentrate on the case of positive characteristic, i.e. of function fields, both because it is technically simpler and also because it allows to keep track of the geometric significance of the construction.

In order to understand how to perform in the global case, the cutoff $R_\Lambda = \widehat{P}_\Lambda P_\Lambda$ of section VIII, we shall first analyze the relative position of the pair of projections

$\widehat{P}_\Lambda$, P_Λ when $\Lambda \to \infty$. Thus, we let $S \supset S_0$ be a finite set of places of k, large enough so that $mod(C_S) = mod(C_k) = q^{\mathbb{Z}}$ and that for any fundamental domain D for the action of O_S^* on J_S, the product $D \times \prod R_v^*$ is a fundamental domain for the action of k^* on J_k.

Both $\widehat{P}_\Lambda$ and P_Λ commute with the decomposition of $L^2(X_S)$ as the direct sum of the subspaces, indexed by characters χ_0 of $C_{S,1}$,

$$(3) \qquad L_{\chi_0}^2 = \{\xi \in L^2(X_S)\,;\, \xi(a^{-1}x) = \chi_0(a)\,\xi(x),\ \forall\, x \in X_S\,, a \in C_{S,1}\}$$

which corresponds to the projections $P_{\chi_0} = \int \overline{\chi_0}(a)\,U(a)\,d_1\,a$, where $d_1\,a$ is the Haar measure of total mass 1 on $C_{S,1}$.

Lemma 1. *Let χ_0 be a character of $C_{S,1}$, then for Λ large enough $\widehat{P}_\Lambda$ and P_Λ commute on the Hilbert space $L_{\chi_0}^2$.*

We can thus rewrite Theorem VIII.3 in the case of positive characteristic as,

Corollary 2. *Let Q_Λ be the orthogonal projection on the subspace of $L^2(X_S)$ spanned by the $f \in \mathcal{S}(A_S)$ which vanish as well as their Fourier transform for $|x| > \Lambda$. Let $h \in \mathcal{S}(C_S)$ have compact support. Then when $\Lambda \to \infty$, one has*

$$\mathrm{Trace}\,(Q_\Lambda\,U(h)) = 2h(1)\log'\Lambda + \sum_{v \in S} \int_{k_v^*}^{\prime} \frac{h(u^{-1})}{|1-u|}\,d^*u + o(1)$$

where $2\log'\Lambda = \int_{\lambda \in C_S,\,|\lambda| \in [\Lambda^{-1},\Lambda]}\,d^*\lambda$, and the other notations are as in Theorem VIII.3

In fact the proof of lemma 1 ([Co]) shows that the subspaces B_Λ stabilize very quickly, so that the natural map $\xi \to \xi \otimes 1_R$ from $L^2(X_S)$ to $L^2(X_S')$ for $S \subset S'$ maps B_Λ^S onto $B_\Lambda^{S'}$.

We thus get from corollary 2 an S-independent global formulation of the cutoff and of the trace formula. We let $L^2(X)$ be the Hilbert space $L_\delta^2(X)$ of section V for the trivial value $\delta = 0$ which of course eliminates the unpleasant term from the inner product, and we let Q_Λ be the orthogonal projection on the subspace B_Λ of $L^2(X)$ spanned by the $f \in \mathcal{S}(A)$ which vanish as well as their Fourier transform for $|x| > \Lambda$. As we mentionned earlier, the proof of lemma 1 shows that for S and Λ large enough (and fixed character χ), the natural map $\xi \to \xi \otimes 1_R$ from $L^2(X_S)_\chi$ to $L^2(X)_\chi$ maps B_Λ^S onto B_Λ.

It is thus natural to expect that the following global analogue of the trace formula of corollary 2 actually holds, i.e. that when $\Lambda \to \infty$, one has,

$$(4) \qquad \mathrm{Trace}\,(Q_\Lambda\,U(h)) = 2h(1)\log'\Lambda + \sum_v \int_{k_v^*}^{\prime} \frac{h(u^{-1})}{|1-u|}\,d^*u + o(1)$$

where $2\log'\Lambda = \int_{\lambda \in C_k,\,|\lambda| \in [\Lambda^{-1},\Lambda]}\,d^*\lambda$, and the other notations are as in Theorem VIII.3.

We can prove directly that (4) holds when h is supported by $C_{k,1}$ but are not able to prove (4) directly for arbitrary h (even though the right hand side of the formula only contains finitely many nonzero terms since $h \in S(C_k)$ has compact support). What we shall show however is that the trace formula (4) implies the positivity of the Weil distribution, and hence the validity of RH for k. Remember that we are still in positive characteristic where RH is actually a theorem of A.Weil. It will thus be important to check the actual equivalence between the validity of RH and the formula (4). This is achieved by,

Theorem 3.([Co]) *Let k be a global field of positive characteritic and Q_Λ be the orthogonal projection on the subspace of $L^2(X)$ spanned by the $f \in S(A)$ such that $f(x)$ and $\widehat{f}(x)$ vanish for $|x| > \Lambda$. Let $h \in S(C_k)$ have compact support. Then the following conditions are equivalent,*

a) When $\Lambda \to \infty$, one has

$$\mathrm{Trace}\,(Q_\Lambda\, U(h)) = 2h(1)\log'\Lambda + \sum_v \int_{k_v^*}' \frac{h(u^{-1})}{|1-u|}\, d^*u + o(1)$$

b) All L functions with Grössencharakter on k satisfy the Riemann Hypothesis.

To prove that a) implies b), one proves (assuming a)) the positivity of the Weil distribution

$$(5) \qquad\qquad \Delta = \log|d^{-1}|\,\delta_1 + D - \sum_v D_v\,.$$

First, by theorem V.2 applied for $\delta = 0$, the map E,

$$(6) \qquad\qquad E(f)\,(g) = |g|^{1/2} \sum_{q \in k^*} f(qg) \qquad \forall\, g \in C_k\,,$$

defines a surjective isometry from $L^2(X)_0$ to $L^2(C_k)$ such that,

$$(7) \qquad\qquad E\,U(a) = |a|^{1/2}\,V(a)\,E\,,$$

where the left regular representation V of C_k on $L^2(C_k)$ is given by,

$$(8) \qquad\qquad (V(a)\,\xi)\,(g) = \xi(a^{-1}\,g) \qquad \forall\, g, a \in C_k\,.$$

Let S_Λ be the subspace of $L^2(C_k)$ given by,

$$(9) \qquad\qquad S_\Lambda = \{\xi \in L^2(C_k)\,;\ \xi(g) = 0,\ \forall\, g\,,\ |g| \notin [\Lambda^{-1}, \Lambda]\}\,.$$

We shall denote by the same letter the corresponding orthogonal projection.

Let $B_{\Lambda,0}$ be the subspace of $L^2(X)_0$ spanned by the $f \in S(A)_0$ such that $f(x)$ and $\widehat{f}(x)$ vanish for $|x| > \Lambda$ and $Q_{\Lambda,0}$ be the corresponding orthogonal projection.

Let $f \in \mathcal{S}(A)_0$ be such that $f(x)$ and $\widehat{f}(x)$ vanish for $|x| > \Lambda$, then $E(f)(g)$ vanishes for $|g| > \Lambda$, and the equality,

$$(10) \qquad\qquad E(f)(g) = E(\widehat{f}) \left(\frac{1}{g}\right) \qquad f \in \mathcal{S}(A)_0,$$

shows that $E(f)(g)$ vanishes for $|g| < \Lambda^{-1}$.

This shows that $E(B_{\Lambda,0}) \subset S_\Lambda$, so that if we let $Q'_{\Lambda,0} = E \, Q_{\Lambda,0} \, E^{-1}$, we get the inequality,

$$(11) \qquad\qquad Q'_{\Lambda,0} \leq S_\Lambda$$

and for any Λ the following distribution on C_k is of positive type,

$$(12) \qquad\qquad \Delta_\Lambda(f) = \mathrm{Trace}\,((S_\Lambda - Q'_{\Lambda,0})\, V(f)),$$

i.e. one has,

$$(13) \qquad\qquad \Delta_\Lambda(f * f^*) \geq 0,$$

where $f^*(g) = \overline{f}(g^{-1})$ for all $g \in C_k$.

Let then $f(g) = |g|^{-1/2} h(g^{-1})$, so that one has $E\,U(h) = V(\tilde{f})E$ where $\tilde{f}(g) = f(g^{-1})$ for all $g \in C_k$. One has,

$$(14) \qquad\qquad \sum_v D_v(f) - \log|d^{-1}| = \sum_v \int_{k_v^*}^{\prime} \frac{h(u^{-1})}{|1-u|} \, d^*u.$$

One has $\mathrm{Trace}\,(S_\Lambda\, V(f)) = 2f(1) \log' \Lambda$, thus using a) we see that the limit of Δ_Λ when $\Lambda \to \infty$ is the Weil distribution Δ (cf. section V). The term D in the latter comes from the nuance between the subspaces B_Λ and $B_{\Lambda,0}$. This shows using (13), that the distribution Δ is of positive type so that b) holds (cf. [W3]).

To show that b) implies a), one computes from the zeros of L-functions and independently of any hypothesis the limit of the distributions Δ_Λ when $\Lambda \to \infty$. We choose (non canonically) an isomorphism

$$(15) \qquad\qquad C_k \simeq C_{k,1} \times N\,.$$

where $N = \mathrm{range}\, |\;| \subset \mathbb{R}_+^*$, $N \simeq \mathbb{Z}$ is the subgroup $q^{\mathbb{Z}} \subset \mathbb{R}_+^*$.

For $\rho \in \mathbb{C}$ we let $d\mu_\rho(z)$ be the harmonic measure of ρ with respect to the line $i\,\mathbb{R} \subset \mathbb{C}$. It is a probability measure on the line $i\,\mathbb{R}$ and coincides with the Dirac mass at ρ when ρ is on the line.

The implication b)$\Rightarrow$a) follows immediately from the explicit formulas and the following lemma,

Lemma 4. *The limit of the distributions Δ_Λ when $\Lambda \to \infty$ is given by,*

$$\Delta_\infty(f) = \sum_{\substack{L\left(\tilde{\chi}, \frac{1}{2}+\rho\right)=0 \\ \rho \in B/N^\perp}} N(\tilde{\chi}, \tfrac{1}{2} + \rho) \int_{z \in i\,\mathbb{R}} \widehat{f}(\tilde{\chi}, z) d\mu_\rho(z)$$

where B is the open strip $B = \{\rho \in \mathbb{C}; Re(\rho) \in]\frac{-1}{2}, \frac{1}{2}[\}$, $N(\tilde{\chi}, \frac{1}{2} + \rho)$ is the multiplicity of the zero, $d\mu_\rho(z)$ is the harmonic measure of ρ with respect to the line $i\mathbb{R} \subset \mathbb{C}$, and the Fourier transform $\hat{f}$ of f is defined by,

$$\hat{f}(\tilde{\chi}, \rho) = \int_{C_k} f(u)\, \tilde{\chi}(u)\, |u|^\rho\, d^* u.$$

One should compare this lemma with Corollary 3 of section V. In the latter only the critical zeros were coming into play and with a multiplicity controlled by δ. In the above lemma, all zeros do appear and with their full multiplicity, but while the critical zeros appear per-se, the non-critical ones play the role of resonances as in the Fermi theory.

Let us now explain how the above results extend to number fields k. We first need to analyze, as above, the relative position of the projections P_Λ and $\hat{P}_\Lambda$. Let us first remind the reader of the well known geometry of pairs of projectors. Recall that a pair of orthogonal projections P_i in Hilbert space is the same thing as a unitary representation of the dihedral group $\Gamma = \mathbb{Z}/2 * \mathbb{Z}/2$. To the generators U_i of Γ correspond the operators $2P_i - 1$. The group Γ is the semidirect product of the subgroup generated by $U = U_1 U_2$ by the group $\mathbb{Z}/2$, acting by $U \mapsto U^{-1}$. Its irreducible unitary representations are parametrized by an angle $\theta \in [0, \frac{\pi}{2}]$, the corresponding orthogonal projections P_i being associated to the one dimensional subspaces $y = 0$ and $y = x\, tg(\theta)$ in the Euclidean x, y plane. In particular these representations are at most two dimensional. A general unitary representation is characterized by the operator Θ whose value is the above angle θ in the irreducible case. It is uniquely defined by the equality,

(16) $\mathrm{Sin}(\Theta) = |P_1 - P_2|,$

and commutes with P_i.

The first obvious difficulty is that when v is an Archimedian place there exists no non-zero function on k_v which vanishes as well as its Fourier transform for $|x| > \Lambda$. This would be a difficult obstacle were it not for the work of Landau, Pollak and Slepian ([LPS]) in the early sixties, motivated by problems of electrical engineering, which allows to overcome it by showing that though the projections P_Λ and $\hat{P}_\Lambda$ do not commute exactly even for large Λ, their angle is sufficiently well behaved so that the subspace B_Λ makes good sense.

For simplicity we shall take $k = \mathbb{Q}$, so that the only infinite place is real. Let P_Λ be the orthogonal projection onto the subspace,

(17) $P_\Lambda = \{\xi \in L^2(\mathbb{R})\,;\ \xi(x) = 0,\ \forall x\,,\ |x| > \Lambda\}\,.$

and $\hat{P}_\Lambda = FP_\Lambda F^{-1}$ where F is the Fourier transform associated to the basic character $\alpha(x) = e^{-2\pi i x}$. What the above authors have done is to analyze the relative position of the projections P_Λ, $\widehat{P}_\Lambda$ for $\Lambda \to \infty$ in order to account for the obvious existence of signals (a recorded music piece for instance) which for all practical purposes have finite support both in the time variable and the dual frequency variable.

The key observation of ([LPS]) is that the following second order differential operator on $\mathbb{R}$ actually commutes with the projections P_Λ, $\widehat{P}_\Lambda$,

(18) $$H_\Lambda \psi(x) = -\partial((\Lambda^2 - x^2)\,\partial)\psi(x) + (2\pi\Lambda x)^2\,\psi(x),$$

where ∂ is ordinary differentiation in one variable. Exactly as the generator $x\,\partial$ of scaling commutes with the orthogonal projection on the space of functions with positive support, the operator $\partial((\Lambda^2 - x^2)\,\partial)$ commutes with P_Λ. Moreover H_Λ commutes with Fourier transform F, and the commutativity of H_Λ with $\widehat{P}_\Lambda$ thus follows.

If one sticks to functions with support in $[-\Lambda, \Lambda]$, the operator H_Λ has discrete simple spectrum, and was studied long before the work of [LPS]. It appears from the factorization of the Helmoltz equation $\Delta \psi + k^2 \psi = 0$ in one of the few separable coordinate systems in Euclidean 3-space, called the prolate spheroidal coordinates. Its eigenvalues $\chi_n(\Lambda), n \geq 0$ are simple and positive. The corresponding eigenfunctions ψ_n are called the prolate spheroidal wave functions and since $P_\Lambda \widehat{P}_\Lambda P_\Lambda$ commutes with H_Λ, they are the eigenfunctions of $P_\Lambda \widehat{P}_\Lambda P_\Lambda$. A lot is known about them, in particular one can take them to be real valued, and they are even for n even and odd for n odd. The key result of [LPS] is that the corresponding eigenvalues λ_n of the operator $P_\Lambda \widehat{P}_\Lambda P_\Lambda$ are decreasing very slowly from $\lambda_0 \simeq 1$ until the value $n \simeq 4\Lambda^2$ of the index n, they then decrease from $\simeq 1$ to $\simeq 0$ in an interval of length $\simeq \log(\Lambda)$ and then stay close to 0. Of course this gives the eigenvalues of Θ, it dictates the analogue of the subspace B_Λ of lemma 1, as the linear span of the ψ_n, $n \leq 4\Lambda^2$, and it gives the justification of the semi-classical counting of the number of quantum mechanical states which are localized in the interval $[-\Lambda, \Lambda]$ as well as their Fourier transform as the area of the corresponding square in phase space.

We now know what is the subspace B_Λ for the single place ∞, and to obtain it for an arbitrary set of places (containing the infinite one), we just use the same rule as in the case of function fields, i.e. we consider the map,

(19) $$\psi \mapsto \psi \otimes 1_R,$$

which suffices when we deal with the Riemann zeta function. Note also that in that case we restrict ourselves to even functions on $\mathbb{R}$. This gives the analogue of Corollary 2, Theorem 3, and Lemma 4.

We refer to [Co] to see how the formula for the number of zeros

(20) $$N(E) \sim (E/2\pi)(\log(E/2\pi) - 1) + 7/8 + o(1) + N_{osc}(E)$$

appears from our spectral interpretation.

The filtration Q_Λ, S_Λ of the short complex,

(21) $$0 \to L^2(X)_0 \to L^2(C_k) \to 0,$$

allows to define Adelic cohomology in which all nontrivial zeros of L-functions do appear and to complete the following dictionary with unproved last line, between the function theory and the geometry of the Adele class space X,

Function Theory	*Geometry*
Zeros and poles of Zeta	Action of C_k on Adelic cohomology
Functional Equation	$*$ operation
Explicit formula	Lefschetz formula for action of C_k on X
Riemann Hypothesis	Trace formula

General remarks.

a) There is a close analogy between the construction of the Hilbert space $L^2(X)$ in section V, and the construction of the physical Hilbert space ([S] theorem 2.1) in constructive quantum field theory, in the case of gauge theories. In both cases the action of the invariance group (the group $k^* = GL_1(k)$ in our case, the gauge group in the case of gauge theories) is wiped out by the very definition of the inner product. Compare with ([S]) top of page 17.

b) It is quite remarkable that the eigenvalues of the angle operator Θ which we discussed above, also play a key role in the theory of random hermitian matrices. To be more specific, let $E(n, s)$ be the large N limit of the probability that there are exactly n eigenvalues of a random Hermitian $N \times N$ matrix in the interval $[-\frac{\pi}{\sqrt{2N}}t, \frac{\pi}{\sqrt{2N}}t]$, $t = s/2$.

Let us compute this probability $E(n, s)$ (cf.[Me]). One has by construction $\sum_n E(n, s) = 1$. We will do the computation for $n = 0$, for other values of n the computation is similar.

With the notations of section III, $E(0, s)$ is clearly given by the large N limit of,

$$\int\limits_{|E_j| \geq \theta} p_N(E_1, \ldots E_N) dE_1 \ldots dE_N,$$

where $\theta = \frac{\pi}{\sqrt{2N}}t$. This equals

$$\frac{1}{N!} \int\limits_{|E_j| \geq \theta} \sum_{\sigma, \pi \in S_N} \epsilon(\sigma)\epsilon(\pi)\Phi_{\sigma(1)}(E_1)\Phi_{\pi(1)}(E_1)\Phi_{\sigma(2)}(E_2)\Phi_{\pi(2)}(E_2)\ldots$$

$$\Phi_{\sigma(N)}(E_N)\Phi_{\pi(N)}(E_N)dE_1 dE_2 \ldots dE_N =,$$

$$\sum_{\sigma} \epsilon(\sigma) \prod_1^N \big(\Phi_1, (1 - P_\theta)\Phi_{\sigma(1)}\big) \ldots \big(\Phi_N, (1 - P_\theta)\Phi_{\sigma(N)}\big)$$

where P_θ is the operator of multiplication by $1_{[-\theta,\theta]}$, the characteristic function of the interval $[-\theta, \theta]$.

We rewrite this as $\det\left(K_N(1 - P_\theta)K_N\right)\big|_{\text{Range of } K_N} = \prod_1^N(1 - \lambda_{j,N})$, where $\lambda_{j,N}$ are the nonzero eigenvalues of $K_N P_\theta$.

For $N \to \infty$, the equality III.14 allows to replace K_N by the operator given by the kernel,

$$k(x, y) = \frac{\sin \pi(x - y)}{\pi(x - y)}$$

Hence we get as $N \to \infty$,

$$E(0,s) = \prod_1^\infty (1 - \lambda_j(s)),$$

where $s = 2t$, and the $\lambda_j(s)$ are the eigenvalues of the operator $\widehat{P_\pi} P_t$. Here we let, as above, $\widehat{P_\lambda} = \mathcal{F} P_\lambda \mathcal{F}^{-1}$, and $\mathcal{F}$ denotes the Fourier transform, $\mathcal{F}\xi(u) = \int e^{ixu} \xi(x) dx$. Note finally that the eigenvalues of $\widehat{P_a} P_b$ only depend upon the product ab so that the relation with the eigenvalues of Θ should be clear.

c) This paper was finalized during my visit to O.S.U. in October - November, 1998 and I am grateful to this University for its warm hospitality and to A. Gorokhovsky for taking careful notes in my class.

Bibliography.

[AB] M.F. Atiyah and R. Bott, A Lefschetz fixed point formula for elliptic complexes: I, *Annals of Math*, **86** (1967), 374-407.

[B] M. Berry, Riemann's zeta function: a model of quantum chaos, *Lecture Notes in Physics*, **263**, Springer (1986).

[Bg] A. Beurling, A closure problem related to the (zeta function, *Proc. Nat. Ac. Sci.* **41** (1955), 312-314.

[BC] J.-B. Bost and A. Connes, Hecke Algebras, Type III factors and phase transitions with spontaneous symmetry breaking in number theory, *Selecta Mathematica, New Series* **1**, No.3 (1995), 411-457.

[BG] O. Bohigas and M. Giannoni, Chaotic motion and random matrix theories, *Lecture Notes in Physics*, **209** (1984), 1-99.

[BK] M. Berry and J. Keating, $H = qp$ and the Riemann zeros, 'Supersymmetry and Trace Formulae: Chaos and Disorder', edited by J.P. Keating, D.E. Khmelnitskii and I.V. Lerner (Plenum Press).

[Br] F. Bruhat, Distributions sur un groupe localement compact et applications á l'étude des représentations des groupes p-adiques.*Bull. Soc. Math. france.* **89** (1961), 43-75.

[Co1] A. Connes, Classification of injective factors, *Ann. of Math.*, **104**, n. 2 (1976), 73-115.

[Co2] A. Connes, Une classification des facteurs de type III, *Ann. Sci. Ecole Norm. Sup.*, **6**, n. 4 (1973), 133-252.

[Co3] A. Connes, Formule de trace en Géométrie non commutative et hypothèse de Riemann, *C.R. Acad. Sci. Paris Ser. A-B* (1996)

[CT] A. Connes and M. Takesaki, The flow of weights on factors of type III, *Tohoku Math. J.*, **29** (1977), 473-575.

[Co] A. Connes, Trace formula in Noncommutative Geometry and the zeros of the Riemann zeta function. To appear in Selecta Mathematica.

[C] A. Connes, Noncommutative Geometry, Academic Press (1994).

[D] C. Deninger, Local L-factors of motives and regularised determinants, *Invent. Math.*, **107** (1992), 135-150.

[G] D. Goldfeld, A spectral interpretation of Weil's explicit formula, *Lecture Notes in Math.*, **1593**, Springer Verlag (1994), 135-152.

[GS] V. Guillemin and S. Sternberg, Geometric asymptotics, *Math. Surveys*, **14**, *Amer. Math. Soc., Providence, R.I.* (1977)

[Gu] V. Guillemin, Lectures on spectral theory of elliptic operators, *Duke Math. J.*, **44**, No.3 (1977), 485-517.

[Ha] U. Haagerup, Connes' bicentralizer problem and uniqueness of the injective factor of type III_1, *Acta Math.*, **158** (1987), 95-148.

[H] S. Haran, Riesz potentials and explicit sums in arithmetic, *Invent. Math.*, **101** (1990), 697-703.

[J] B. Julia, Statistical theory of numbers, *Number Theory and Physics, Springer Proceedings in Physics*, **47** (1990).

[Ka] M. Kac, Statistical Independence in Probability, *Analysis and Number Theory, Carus Math. Monographs* **18** (1959).

[KS] N. Katz and P. Sarnak, Random matrices, Frobenius eigenvalues and Monodromy, (1996) , Book, to appear.

[KS] N. Katz and P. Sarnak, Zeros of zeta functions, their spacings and spectral nature, (1997), to appear.

[K] W. Krieger, On ergodic flows and the isomorphism of factors, *Math. Ann.*, **223** (1976), 19-70.

[Laca] M. Laca, From Endomorphisms to Automorphisms and back, Dilations and Full Corners, *preprint*. (1998).

[LPS1] D. Slepian and H. Pollak, Prolate spheroidal wave functions, Fourier analysis and uncertainty I, *Bell Syst. Tech. J.* **40** (1961).

[LPS2] H.J. Landau and H. Pollak, Prolate spheroidal wave functions, Fourier analysis and uncertainty II, *Bell Syst. Tech. J.* **40** (1961).

[LPS3] H.J. Landau and H. Pollak, Prolate spheroidal wave functions, Fourier analysis and uncertainty III, *Bell Syst. Tech. J.* **41** (1962).

[M] H. Montgomery, The pair correlation of zeros of the zeta function, *Analytic Number Theory*, AMS (1973).

[Me] M.L. Mehta, Random matrices, Academic Press,(1991).

[Mo] G. Moore, Arithmetic and Attractors, hep-th/9807087.

[O] A. Odlyzko, On the distribution of spacings between zeros of zeta functions, *Math. Comp.* **48** (1987), 273-308.

[P] G. Pólya, Collected Papers, Cambridge, M.I.T. Press (1974).

[Pat] S. Patterson, An introduction to the theory of the Riemann zeta function, *Cambridge Studies in advanced mathematics*, **14** Cambridge University Press (1988).

[R] B. Riemann, Mathematical Werke, Dover, New York (1953).

[S] E.Seiler, Gauge Theories as a problem of constructive Quantum Field Theory and Statistical Mechanics, Lecture Notes in Physics **159** Springer (1982).

[Se] A.Selberg, *Collected papers*, Springer (1989).

[T] M. Takesaki, *Tomita's theory of modular Hilbert algebras and its applications*, Lecture Notes in Math. bf 128, Springer (1989).

[Ta] M. Takesaki, Duality for crossed products and the structure of von Neumann algebras of type III, *Acta Math.* **131** (1973), 249-310.

[W1] A. Weil, Basic Number Theory, Springer, New York (1974).

[W2] A. Weil, Fonctions zêta et distributions, *Séminaire Bourbaki*, **312**, (1966).

[W3] A. Weil, Sur les formules explicites de la théorie des nombres, *Izv. Mat. Nauk.*, (Ser. Mat.) **36**, 3-18.

[W4] A. Weil, Sur la théorie du corps de classes,*J. Math. Soc. Japan*, **3**, (1951).

[W5] A. Weil, Sur certains groupes d'operateurs unitaires,*Acta Math.* , **111**, (1964).

 [Z] D. Zagier, Eisenstein series and the Riemannian zeta function, *Automorphic Forms, Representation Theory and Arithmetic,* Tata, Bombay (1979), 275-301.

Partial Differential Equations and Monge–Kantorovich Mass Transfer

Lawrence C. Evans*
Department of Mathematics
University of California, Berkeley

1. **Introduction**
   1.1 Optimal mass transfer
   1.2 Relaxation, duality

   Part I: Cost $= \frac{1}{2}(\text{Distance})^2$

2. **Heuristics**
   2.1 Geometry of optimal transport
   2.2 Lagrange multipliers
3. **Optimal mass transport, polar factorization**
   3.1 Solution of dual problem
   3.2 Existence of optimal mass transfer plan
   3.3 Polar factorization of vector fields
4. **Regularity**
   4.1 Solving the Monge–Ampere equation
   4.2 Examples
   4.3 Interior regularity for convex targets
   4.4 Boundary regularity for convex domain and target
5. **Application: Nonlinear interpolation**
6. **Application: Time-step minimization and nonlinear diffusion**
   6.1 Discrete time approximation
   6.2 Euler–Lagrange equation
   6.3 Convergence
7. **Application: Semigeostrophic models in meteorology**
   7.1 The PDE in physical variables
   7.2 The PDE in dual variables
   7.3 Frontogenesis

*Supported in part by NSF Grant DMS-94-24342. This paper will appear in Current Developments in Mathematics 1997.

Part II: Cost = Distance

8. **Heuristics**
   8.1 Geometry of optimal transport
   8.2 Lagrange multipliers
9. **Optimal mass transport**
   9.1 Solution of dual problem
   9.2 Existence of optimal mass transfer plan
   9.3 Detailed mass balance, transport density
10. **Application: Shape optimization**
11. **Application: Sandpile models**
    11.1 Growing sandpiles
    11.2 Collapsing sandpiles
    11.3 A stochastic model
12. **Application: Compression molding**

Part III: Appendix

13. **Finite-dimensional linear programming**

References

In Memory of
Frederick J. Almgren, Jr.
and
Eugene Fabes

1 Introduction

These notes are a survey documenting an interesting recent trend within the calculus of variations, the rise of differential equations techniques for Monge–Kantorovich type optimal mass transfer problems. I will discuss in some detail a number of recent papers on various aspects of this general subject, describing newly found applications in the calculus of variations itself and in physics. An important theme will be the rather different analytic and geometric tools for, and physical interpretations of, Monge–Kantorovich problems with a uniformly convex cost density (here exemplified by $c(x,y) = \frac{1}{2}|x - y|^2$) versus those problems with a nonuniformly convex cost (exemplified by $c(x,y) = |x - y|$). We will as well study as applications several physical processes evolving in time, for which we can identify optimal Monge–Kantorovich mass transferences on "fast" time scales.

1.1 Optimal mass transfer

The original transport problem, proposed by Monge in the 1780's, asks how best to move a pile of soil or rubble ("déblais") to an excavation or fill ("remblais"), with the least amount of work. In modern parlance, we are given two nonnegative Radon measures $\mu^{\pm}$ on $\mathbb{R}^n$, satisfying the overall *mass balance* condition

$$\mu^+(\mathbb{R}^n) = \mu^-(\mathbb{R}^n) < \infty, \qquad (1.1)$$

and we consider the class of measurable, one-to-one mappings $\mathbf{s} : \mathbb{R}^n \to \mathbb{R}^n$ which rearrange μ^+ into μ^-:

$$\mathbf{s}_{\#}(\mu^+) = \mu^-. \qquad (1.2)$$

In other words, we require

$$\int_X h(\mathbf{s}(x))\, d\mu^+(x) = \int_Y h(y)\, d\mu^-(y) \qquad (1.3)$$

for all continuous functions h, where $X = \mathrm{spt}(\mu^+)$, $Y = \mathrm{spt}(\mu^-)$. We denote by $\mathcal{A}$ the admissible class of mappings $\mathbf{s}$ as above, satisfying (1.2), (1.3)

Given also is the *work* or *cost density* function

$$c : \mathbb{R}^n \times \mathbb{R}^n \to [0, \infty);$$

so that $c(x,y)$ records the work required to move a unit mass from the position $x \in \mathbb{R}^n$ to a new position $y \in \mathbb{R}^n$. (In Monge's original problem $c(x, y) = |x - y|$; that is, the work is simply proportional to the distance moved.)

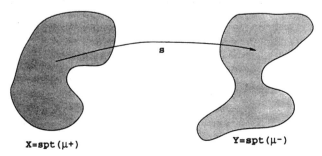

X=spt (μ+)   Y=spt (μ-)

The total *work* corresponding to a mass rearrangement plan $\mathbf{s} \in \mathcal{A}$ is thus

$$I[\mathbf{s}] := \int_{\mathbb{R}^n} c(x, \mathbf{s}(x)) \, d\mu^+(x). \qquad (1.4)$$

Our problem is therefore to find and characterize an *optimal mass transfer* $\mathbf{s}^* \in \mathcal{A}$ which minimizes the work:

$$I[\mathbf{s}^*] = \min_{\mathbf{s} \in \mathcal{A}} I[\mathbf{s}]. \qquad (1.5)$$

In other words, we wish to construct a one-to-one mapping $\mathbf{s}^* : \mathbb{R}^n \to \mathbb{R}^n$ which pushes the measure μ^+ onto μ^- and, among all such mappings, minimizes $I[\cdot]$. We will later see that a really remarkable array of interesting mathematical and physical interpretations follow.

This is even now, over two hundred years later, a difficult mathematical problem, owing mostly to the highly nonlinear structure of the constraint. For instance, if $\mu^\pm$ have smooth densities $f^\pm$, that is, if

$$d\mu^+ = f^+ dx, \ d\mu^- = f^- dy, \qquad (1.6)$$

then (1.2) reads

$$f^+(x) = f^-(\mathbf{s}(x)) \det(D\mathbf{s}(x)) \quad (x \in X), \qquad (1.7)$$

where we write $\mathbf{s} = (s^1, \ldots, s^n)$ and

$$D\mathbf{s} = \begin{pmatrix} s^1_{x_1} & \cdots & s^1_{x_n} \\ & \ddots & \\ s^n_{x_1} & \cdots & s^n_{x_n} \end{pmatrix}_{n \times n} = \text{Jacobian matrix of the mapping } \mathbf{s}.$$

It is not at all apparent offhand that there exists any mapping, much less an optimal mapping, satisfying this constraint. Additionally, if $\{\mathbf{s}_k\}_{k=1}^\infty \subset \mathcal{A}$ is a minimizing sequence,

$$I[\mathbf{s}_k] \to \inf_{\mathbf{s}\in\mathcal{A}} I[\mathbf{s}],$$

an obvious guess is that we can somehow pass to a subsequence $\{\mathbf{s}_{k_j}\}_{j=1}^\infty \subset \{\mathbf{s}_k\}_{k=1}^\infty$, which in turn converges to an optimal mass allocation plan $\mathbf{s}^*$:

$$\mathbf{s}_{k_j} \to \mathbf{s}^*.$$

However, there is no clear way to extract such a subsequence, converging in any reasonable sense to a limit. The direct methods of the calculus of variations fail spectacularly, as there are no terms creating any sort of compactness built into the work functional $I[\cdot]$. In particular, $I[\cdot]$ does not involve the gradient of $\mathbf{s}$ at all and so is not coercive on any Sobolev space. And yet, as we will momentarily see, precisely this feature opens up the problem to methods of linear programming.

1.2 Relaxation, duality

Kantorovich in the 1940's [K1],[K2] (see also [R]) resolved certain of these problems by introducing:

(i) a "relaxed" variant of Monge's original mass allocation problem and, more importantly,

(ii) a dual variational principle.

The idea behind (i) is, remarkably, to transform (1.5) into a linear problem. The trick is firstly to introduce the class

$$\mathcal{M} := \left\{ \text{Radon probability measures } \mu \text{ on } \mathbb{R}^n \times \mathbb{R}^n \mid \text{proj}_x\mu = \mu^+, \ \text{proj}_y\mu = \mu^- \right\} \tag{1.8}$$

of measures on the product space $\mathbb{R}^n \times \mathbb{R}^n$, whose projections on the first n coordinates and the last n coordinates are, respectively, μ^+, μ^-. Given then $\mu \in \mathcal{M}$, we define the *relaxed cost* functional

$$J[\mu] := \int_{\mathbb{R}^n \times \mathbb{R}^n} c(x,y)\, d\mu(x,y). \tag{1.9}$$

The point of course is that if we have a mapping $\mathbf{s} \in \mathcal{A}$, then the induced measure

$$\mu(E) := \mu^+\{x \in \mathbb{R}^n \mid (x, \mathbf{s}(x)) \in E\} \quad (E \subset \mathbb{R}^n \times \mathbb{R}^n, \ E \text{ Borel}) \tag{1.10}$$

belongs to $\mathcal{M}$. Furthermore the new functional (1.9) is *linear* in μ and so, under appropriate assumptions on the cost c, simple compactness arguments assert the existence of at least one optimal measure $\mu^* \in \mathcal{M}$, satisfying

$$J[\mu^*] = \min_{\mu \in \mathcal{M}} J[\mu]. \qquad (1.11)$$

Such a measure μ^* need not, however, be generated by any one-to-one mapping $\mathbf{s} \in \mathcal{A}$, and consequently the foregoing construction allows us only to establish the existence of a "weak" or "generalized" solution of Monge's original problem. We will several times later return to the central problem of fashioning some sort of "strong" solution, which actually corresponds to a mapping.

Of even greater importance for our purposes was Kantorovich's additional introduction of a *dual problem*. The best way to motivate this is by analogy with the finite dimensional case. Suppose that then c_{ij}, μ_i^+, μ_j^- $(i = 1, \ldots, n;\ j = 1, \ldots, m)$ are given nonnegative numbers, satisfying the balance condition

$$\sum_{i=1}^{n} \mu_i^+ = \sum_{j=1}^{m} \mu_j^-,$$

and we wish to find μ_{ij}^* $(i = 1, \ldots, n;\ j = 1, \ldots, m)$ to

$$\text{minimize} \quad \sum_{i=1}^{n} \sum_{j=1}^{m} c_{ij} \mu_{ij}, \qquad (1.12)$$

subject to the constraints

$$\sum_{j=1}^{m} \mu_{ij} = \mu_i^+, \ \sum_{i=1}^{n} \mu_{ij} = \mu_j^-, \ \mu_{ij} \geq 0 \quad (i = 1, \ldots, n;\ j = 1, \ldots, m). \qquad (1.13)$$

This is clearly the discrete analogue of (1.8), (1.9), (1.11). As explained in the appendix (§13) the discrete linear programming dual problem to (1.12) is to find $u = (u_1, \ldots, u_n) \in \mathbb{R}^n$, $v = (v_1, \ldots, v_m) \in \mathbb{R}^m$ so as to

$$\text{maximize} \quad \sum_{i=1}^{n} u_i \mu_i^+ + \sum_{j=1}^{m} v_j \mu_j^-, \qquad (1.14)$$

subject to the inequalities

$$u_i + v_j \leq c_{ij}, \ u_i \geq 0, \ v_j \geq 0 \quad (i = 1, \ldots, n;\ j = 1, \ldots, m). \qquad (1.15)$$

We can now by analogy guess the dual variational principle to (1.11). For this, we introduce a continuous variant of (1.15) by defining

$$\mathcal{L} := \{(u, v) \mid u, v : \mathbb{R}^n \to \mathbb{R} \text{ continuous, } u(x) + v(y) \leq c(x, y) \ (x, y \in \mathbb{R}^n)\}. \tag{1.16}$$

(It turns out to be inappropriate to ask as well that $u, v \geq 0$.) Likewise, we introduce the continuous analogue of (1.14) by setting

$$K[u, v] := \int_{\mathbb{R}^n} u(x) \, d\mu^+(x) + \int_{\mathbb{R}^n} v(y) \, d\mu^-(y). \tag{1.17}$$

Consequently our *dual problem* is to find an optimal pair $(u^*, v^*) \in \mathcal{L}$ such that

$$K[u^*, v^*] = \max_{(u,v) \in \mathcal{L}} K[u, v]. \tag{1.18}$$

In summary then, the transformation of Monge's original mass allocation problem (1.5) into the dual problem (1.18) presents us with a rather different vantage point: rather than struggling to construct an optimal mapping $s^* \in \mathcal{A}$ satisfying a highly nonlinear constraint, we are now confronted with the task of finding an optimal pair $(u^*, v^*) \in \mathcal{L}$. And this, as we will see later, is really easy. The mathematical structure of the dual problem provides precisely what was missing for the original problem, enough compactness to construct a minimizer as some sort of limit of a minimizing sequence.

And yet this of course is not the story's end. Kantorovich's methods of first relaxing and then dualizing have brought us into a realm where routine mathematical tools work, but have also taken us far away from the original issue: namely, how do we actually fashion an optimal allocation plan s^*?

We devote much of the remainder of the paper to answering this question, in two most important cases of the uniformly convex cost density:

$$c(x, y) = \frac{1}{2}|x - y|^2 \quad (x, y \in \mathbb{R}^n), \tag{1.19}$$

and the nonuniformly convex cost density

$$c(x, y) = |x - y| \quad (x, y \in \mathbb{R}^n). \tag{1.20}$$

These two intensely studied choices are rich in mathematical structure, and serve as archetypes for other models. Observe carefully the very different geometric consequences: in the first case the graph of the mapping $y \mapsto c(x, y)$ contains no straight lines, and in the second case it does. The latter degeneracy will create interesting problems.

Remark. I make no attempt in this paper to survey the vast literature on Monge–Kantorovich methods in probability and statistics, a nice summary of which may be found in Rachev [R].   □

Part I: Cost $= \frac{1}{2}$(Distance)^2

2 Heuristics

For this and the next five sections we take the quadratic cost density

$$c(x, y) := \frac{1}{2}|x - y|^2 \quad (x, y \in \mathbb{R}^n), \tag{2.1}$$

$|\cdot|$ denoting the usual Euclidean norm. We hereafter wish to understand if and how we can construct an optimal mass transfer plan $\mathbf{s}^*$ solving (1.5), where now

$$I[\mathbf{s}] := \frac{1}{2} \int_{\mathbb{R}^n} |x - \mathbf{s}(x)|^2 \, d\mu^+(x) \tag{2.2}$$

for $\mathbf{s}$ in $\mathcal{A}$, the admissible class of measurable, one-to-one maps of $\mathbb{R}^n$ which push forward the measure μ^+ to μ^-. The following techniques were largely pioneered by Y. Brenier in [B2].

2.1 Geometry of optimal transport

We begin with some informal insights, our goal being to understand, for the moment without proofs, what information about an optimal mapping we can extract directly from the original and dual variational principles. In other words, how can we exploit the very fact that a given mapping minimizes the work functional $I[\cdot]$ (among all other mappings in $\mathcal{A}$), to understand its precise geometric properties?

So now let us assume that $\mathbf{s}^* \in \mathcal{A}$ minimizes the work functional (2.2), among all other mappings $\mathbf{s} \in \mathcal{A}$. Fix a positive integer m, take distinct points $\{x_k\}_{k=1}^m \subset X = \text{spt}\,(\mu^+)$, and assume we can find small disjoint balls

$$E_k := B(x_k, r_k) \quad (k = 1, \ldots, m), \tag{2.3}$$

and radii $\{r_k\}_{k=1}^m$, adjusted so that

$$\mu^+(E_1) = \cdots = \mu^+(E_m) = \varepsilon. \qquad (2.4)$$

Next write $y_k := \mathbf{s}^*(x_k)$, $F_k := \mathbf{s}^*(E_k)$. Since $\mathbf{s}^*$ pushes μ^+ to μ^-, we have

$$\mu^-(F_1) = \cdots = \mu^-(F_m) = \varepsilon. \qquad (2.5)$$

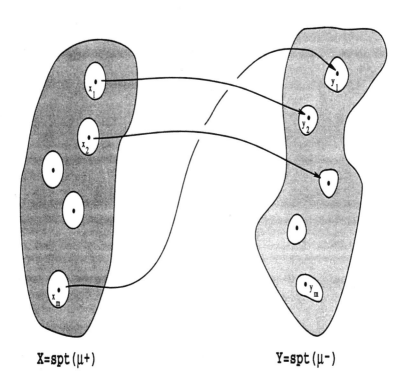

X=spt(μ+)   Y=spt(μ-)

We construct another mapping $\mathbf{s} \in \mathcal{A}$ by cyclically permuting the images of the balls $\{E_k\}_{k=1}^m$. That is, we design $\mathbf{s} \in \mathcal{A}$ so that

$$
\begin{cases}
\mathbf{s}(x_k) = y_{k+1}, \ \mathbf{s}(E_k) = F_{k+1} & (k = 1, \ldots, m) \\
\mathbf{s} \equiv \mathbf{s}^* \text{ on } X - \bigcup_{k=1}^m E_k,
\end{cases}
\tag{2.6}
$$

where $y_{m+1} := y_1$, $F_{m+1} := F_1$.

Then, since $\mathbf{s}^*$ is a minimizer,

$$
I[\mathbf{s}^*] \le I[\mathbf{s}]. \tag{2.7}
$$

Remembering (2.2), we deduce

$$
\sum_{k=1}^m \int_{E_k} |x - \mathbf{s}^*(x)|^2 \, d\mu^+(x) \le \sum_{k=1}^m \int_{E_k} |x - \mathbf{s}(x)|^2 \, d\mu^+(x).
$$

Since both $\mathbf{s}^*, \mathbf{s}$ push μ^+ to μ^-, we can further simplify and then divide by ε:

$$
\sum_{k=1}^m \fint_{E_k} x \cdot (\mathbf{s}(x) - \mathbf{s}^*(x)) \, d\mu^+(x) \le 0, \tag{2.8}
$$

the slash through the integral denoting an average. Now send $\varepsilon \to 0$. Assuming that the mapping $\mathbf{s}^*$ and the measure μ^+ are well-behaved, we deduce from (2.8) that

$$
\sum_{k=1}^m x_k \cdot (y_{k+1} - y_k) \le 0 \tag{2.9}
$$

for $y_k = \mathbf{s}^*(x_k)$, $y_{m+1} = y_1$, $k = 1, 2, \ldots, m$.

In the terminology of convex analysis, (2.9) asserts that the graph $\{(x, \mathbf{s}^*(x)) \mid x \in X\} \subset \mathbb{R}^n \times \mathbb{R}^n$ is *cyclically monotone*. This is an interesting deduction in light of an important theorem of Rockafeller [Rk], asserting that *a cyclically monotone subset of $\mathbb{R}^n \times \mathbb{R}^n$ lies in the subdifferential of a convex mapping of $\mathbb{R}^n$ into $\mathbb{R}$*. In other words,

$$
\mathbf{s}^* \subset \partial \phi^*, \tag{2.10}
$$

for some convex function ϕ^*, in the sense of possibly multivalued graphs on $\mathbb{R}^n \times \mathbb{R}^n$. Moreover, a convex function is differentiable a.e., and so

$$
\mathbf{s}^* = D\phi^* \quad \text{a.e. in } X, \tag{2.11}
$$

where D denotes the gradient.

We have come upon an important deduction: *an optimal mass allocation plan is the gradient of a convex potential ϕ^*.* (Cf. McCann [MC2], etc.)

2.2 Lagrange multipliers

In view of its importance we provide next an alternative, but still strictly formal, analytic derivation of (2.11).

For this, we assume the measures $\mu^\pm$ have smooth densities, $d\mu^+ = f^+dx$, $d\mu^- = f^-dy$, and introduce the *augmented work functional*

$$\tilde{I}[\mathbf{s}] := \int_{\mathbb{R}^n} \frac{1}{2}|x - \mathbf{s}(x)|^2 f^+(x) + \lambda(x)[f^-(\mathbf{s}(x))\det(D\mathbf{s}(x)) - f^+(x)]dx, \tag{2.12}$$

where the function λ is the *Lagrange multiplier* corresponding to the pointwise constraint that $\mathbf{s}_\#(\mu^+) = \mu^-$ (that is, $f^+ = f^-(\mathbf{s})\det(D\mathbf{s})$). Computing the first variation, we find for $k = 1, \ldots, m$:

$$(\lambda f^-(\mathbf{s}^*)(\text{cof}D\mathbf{s}^*)_i^k)_{x_i} = (s^{*k} - x_k)f^+ + \lambda f_{y_k}^-(\mathbf{s}^*)\det(D\mathbf{s}^*). \tag{2.13}$$

Here $\text{cof}D\mathbf{s}^*$ is the cofactor matrix of $D\mathbf{s}^*$; that is, the $(k,i)^{th}$ entry of $\text{cof}D\mathbf{s}^*$ is $(-1)^{k+i}$ times the $(k,i)^{th}$ minor of the matrix $D\mathbf{s}^*$.

Standard matrix identities assert $(\text{cof}D\mathbf{s}^*)_{i,x_i}^k = 0$, $s_{x_i}^{*l}(\text{cof}D\mathbf{s}^*)_i^k = \delta_{kl}(\det D\mathbf{s}^*)$, and $s_{x_j}^{*k}(\text{cof}D\mathbf{s}^*)_i^k = \delta_{ij}(\det D\mathbf{s}^*)$. We employ these equalities to simplify (2.13), and thereby discover after some calculations

$$\lambda_{x_i}f^-(\mathbf{s}^*)(\text{cof}D\mathbf{s}^*)_i^k = (s^{*k} - x_k)f^+. \tag{2.14}$$

Now multiply by $s_{x_j}^{*k}$ and sum on k, to deduce:

$$\lambda_{x_j} = (s^{*k} - x_k)s_{x_j}^{*k}.$$

But then

$$\left(\lambda - \frac{|\mathbf{s}^* - x|^2}{2} + \frac{|x|^2}{2}\right)_{x_j} = s^{*j} \quad (j = 1, \ldots, n),$$

and so (2.11) again follows, for the potential $\phi^* := \lambda - \frac{|\mathbf{s}^* - x|^2}{2} + \frac{|x|^2}{2}$.

3 Optimal mass transport, polar factorization

3.1 Solution of dual problem

The foregoing heuristics done with, we turn next to the task of proving rigorously the existence of an optimal mass allocation plan. We expect

$\mathbf{s}^* = D\phi^*$ almost everywhere for some convex potential ϕ^*, and the task is now really to deduce this. We will do so from the Kantorovich *dual variational principle* (1.16)– (1.18) introduced in §1. We hereafter concentrate on the situation

$$d\mu^+ = f^+ dx, \ d\mu^- = f^- dy, \qquad (3.1)$$

where $f^{\pm}$ are bounded, nonnegative functions with compact support, satisfying the *mass balance* condition

$$\int_X f^+(x)dx = \int_Y f^-(y)\,dy \qquad (3.2)$$

where, as always, $X := \operatorname{spt}(f^+)$, $Y := \operatorname{spt}(f^-)$. For the case at hand, the dual problem is to find (u^*, v^*) so as to *maximize*

$$K[u,v] := \int_X u(x)f^+(x)\,dx + \int_Y v(y)f^-(y)\,dy, \qquad (3.3)$$

subject to the constraint

$$u(x) + v(y) \le \frac{1}{2}|x - y|^2 \quad (x \in X, \ y \in Y). \qquad (3.4)$$

We wish to take up tools from convex analysis, and for this must first change variables:

$$\begin{cases} \phi(x) := \frac{1}{2}|x|^2 - u(x) & (x \in X) \\ \psi(y) := \frac{1}{2}|y|^2 - v(y) & (y \in Y). \end{cases} \qquad (3.5)$$

Note that now (3.4) says

$$\phi(x) + \psi(y) \ge x \cdot y \quad (x \in X, \ y \in Y), \qquad (3.6)$$

and so the variational problem is then to *minimize*

$$L[\phi, \psi] := \int_X \phi(x)f^+(x)\,dx + \int_Y \psi(y)f^-(y)\,dy, \qquad (3.7)$$

subject to the constraint (3.6).

Lemma 3.1 (i) *There exist (ϕ^*, ψ^*) solving this minimization problem.*
(ii) *Furthermore, (ϕ^*, ψ^*) are dual convex functions, in the sense that*

$$\begin{cases} \phi^*(x) = \max_{y \in Y}(x \cdot y - \psi^*(y)) & (x \in X) \\ \psi^*(y) = \max_{x \in X}(x \cdot y - \phi^*(x)) & (y \in Y). \end{cases} \qquad (3.8)$$

Proof. 1. If ϕ, ψ satisfy (3.6), then

$$\phi(x) \geq \max_{y \in Y}(x \cdot y - \psi(y)) =: \hat{\phi}(x) \tag{3.9}$$

and

$$\hat{\phi}(x) + \psi(y) \geq x \cdot y \quad (x \in X, \ y \in Y). \tag{3.10}$$

Consequently

$$\psi(y) \geq \max_{x \in X}(x \cdot y - \hat{\phi}(x)) =: \hat{\psi}(y) \tag{3.11}$$

and

$$\hat{\phi}(x) + \hat{\psi}(y) \geq x \cdot y \quad (x \in X, \ y \in Y). \tag{3.12}$$

As $\psi \geq \hat{\psi}$, (3.9) implies

$$\max_{y \in Y}(x \cdot y - \hat{\psi}(y)) \geq \hat{\phi}(x).$$

This and (3.12) say

$$\hat{\phi}(x) = \max_{y \in Y}(x \cdot y - \hat{\psi}(y)). \tag{3.13}$$

Since $f^{\pm} \geq 0$ and $\psi \geq \hat{\psi}$, $\phi \geq \hat{\phi}$, we see that $L[\hat{\phi}, \hat{\psi}] \geq L[\phi, \psi]$.

2. Consequently in seeking for minimizers of L we may restrict attention to convex dual pairs $(\hat{\phi}, \hat{\psi})$, as above. Such functions are uniformly Lipschitz continuous, and so, after adding or subtracting constants, we can extract a uniformly convergent subsequence from any minimizing sequence for L. We thereby secure an optimal, convex dual pair. □

3.2 Existence of optimal mass transfer plan

Let us now regard (3.8) as defining $\phi^*(x), \psi^*(y)$ for all $x, y \in \mathbb{R}^n$. Then $\phi^*, \psi^* : \mathbb{R}^n \to \mathbb{R}$ are convex, and consequently differentiable a.e. We demonstrate next that

$$\mathbf{s}^*(x) := D\phi^*(x) \quad (\text{a.e. } x \in X) \tag{3.14}$$

solves the mass allocation problem.

Theorem 3.1 *Define* **s*** *by (3.14). Then*
   (i) **s*** : $X \to Y$ *is essentially one-to-one and onto.*
   (ii) $\int_X h(\mathbf{s}^*(x))\,d\mu^+(x) = \int_Y h(y)\,d\mu^-(y)$ *for each* $h \in C(Y)$.
   (iii) *Lastly,*

$$\frac{1}{2}\int_X |x - \mathbf{s}^*(x)|^2\,d\mu^+(x) \le \frac{1}{2}\int_X |x - \mathbf{s}(x)|^2\,d\mu^+(x)$$

for all $\mathbf{s} : X \to Y$ *such that* $\mathbf{s}_\#(\mu^+) = \mu^-$.

Proof. 1. From the max-representation function (3.8) we see that $\mathbf{s}^* = D\phi^* \in Y$ a.e.
   2. Fix $\tau > 0$, and define the variations

$$\begin{cases} \psi_\tau(y) := \psi^*(y) + \tau h(y) & (y \in Y) \\ \phi_\tau(x) := \max_{y \in Y}(x \cdot y - \psi_\tau(y)) & (x \in X). \end{cases} \tag{3.15}$$

Then

$$\phi_\tau(x) + \psi_\tau(y) \ge x \cdot y \quad (x \in X,\ y \in Y) \tag{3.16}$$

and so

$$L[\phi^*, \psi^*] \le L[\phi_\tau, \psi_\tau] =: i(\tau).$$

As the mapping $\tau \mapsto i(\tau)$ thus has a minimum at $\tau = 0$,

$$\begin{aligned} 0 &\le \frac{1}{\tau}(L[\phi_\tau, \psi_\tau] - L[\phi^*, \psi^*]) \\ &= \int_X \left[\frac{\phi_\tau(x) - \phi^*(x)}{\tau}\right] f^+(x)\,dx + \int_Y h(y)f^-(y)\,dy. \end{aligned} \tag{3.17}$$

Now $\left|\frac{\phi_\tau - \phi^*}{\tau}\right| \le \|h\|_{L^\infty}$. Furthermore if we take $y_\tau \in Y$ so that

$$\phi_\tau(x) = x \cdot y_\tau - \psi_\tau(y_\tau), \tag{3.18}$$

then

$$\phi_\tau(x) - \phi^*(x) = x \cdot y_\tau - \psi^*(y_\tau) - \tau h(y_\tau) - \phi^*(x) \le -\tau h(y_\tau). \tag{3.19}$$

On the other hand, if we select $y \in Y$ such that

$$\phi^*(x) = x \cdot y - \psi^*(y), \tag{3.20}$$

then

$$\phi_\tau(x) - \phi^*(x) \geq x \cdot y^* - \psi^*(y) - \tau h(y) - \phi^*(x) = -\tau h(y).$$
(3.21)

Thus

$$-h(y) \leq \frac{\phi_\tau(x) - \phi^*(x)}{\tau} \leq -h(y_\tau).$$
(3.22)

If we take a point $x \in X$ where $s^*(x) := D\phi^*(x)$ exists, then (3.20) implies $y = s^*(x)$. Furthermore as $\tau \to 0$, $y_\tau \to s^*(x)$. Thus (3.17), (3.22) and the Dominated Convergence Theorem imply

$$\int_X h(s^*(x))f^+(x)\,dx \leq \int_Y h(y)f^-(y)\,dy.$$

Replacing h by $-h$, we deduce that equality holds. This is statement (ii).

3. Now take s to be any admissible mapping. Then

$$0 = \int_Y [\psi^*(y) - \psi^*(y)]f^-(y)\,dy = \int_X [\psi^*(s(x)) - \psi^*(s^*(x))]f^+(x)\,dx.$$

Since $\phi^*(x) + \psi^*(y) \geq x \cdot y$, with equality for $y = s(x)$, we have

$$\begin{aligned}
0 &\geq \int_X [x \cdot (s(x) - s^*(x)) - \phi^*(x) + \phi^*(x)]f^+(x)\,dx \\
&= \int_X [x \cdot (s(x) - s^*(x))]f^+(x)\,dx.
\end{aligned}$$

This implies assertion (iii) of the Theorem: s^* is optimal. □

This proof follows Gangbo [G] and Caffarelli [C4]. Another neat approach is due to McCann [MC2]; he approximates the measures $\mu^\pm$ by point masses, solves the resulting discrete linear programming problem, and the passes to limits. See also Gangbo–McCann [G-M1], [G-M2], McCann [MC4]. An interesting related work is Wolfson [W].

3.3 Polar factorization of vector fields

Assume next that $U \subset \mathbb{R}^n$ is open, bounded, with $|\partial U| = 0$, and that $r : U \to \mathbb{R}^n$ is a bounded measurable mapping satisfying the *nondegeneracy condition*

$$\begin{cases} |r^{-1}(N)| = 0 \text{ for each bounded Borel} \\ \text{set } N \subset \mathbb{R}^n, \text{ with } |N| = 0. \end{cases}$$
(3.23)

Define the modified functional

$$\hat{L}[\phi, \psi] := \int_U \phi(\mathbf{r}(y)) + \psi(y)\, dy, \qquad (3.24)$$

which we propose to minimize among functions (ϕ, ψ) satisfying

$$\phi(x) + \psi(y) \geq x \cdot y. \qquad (3.25)$$

Let (ϕ^*, ψ^*) solve this problem. Then taking variations as in the previous proof, we deduce

$$\int_U h(y)\, dy = \int_U h(D\phi^*(\mathbf{r}(y)))\, dy$$

for all $h \in C(U)$. Define now

$$\mathbf{s}^*(x) := D\phi^*(\mathbf{r}(x)).$$

Then $\mathbf{s}^*$ is measure preserving and

$$\mathbf{r}(x) = D\psi^*(\mathbf{s}^*(x)). \qquad (3.26)$$

This is the *polar factorization* of the vector field $\mathbf{r}$, as *the composition of the gradient of a convex function and a measure preserving mapping.*

□

This remarkable polar factorization was established by Brenier [B1], [B2], and the proof later simplified by Gangbo [G]. (Brenier was motivated by problems in fluid mechanics, which we will not discuss in this paper: see [B3].)

Remark. We can regard (3.26) as a nonlinear generalization of the Helmholtz decomposition of a vector field into the sum of a gradient and a divergence-free field. To see this formally, let $\mathbf{a}$ be a given vector field and write

$$\mathbf{r_0} := id, \ \mathbf{s_0} := id, \ \psi_0 := \frac{|x|^2}{2}.$$

Set

$$\mathbf{r}(\tau) := \mathbf{r_0} + \tau \mathbf{a} \qquad (3.27)$$

for small $|\tau|$. The polar factorization gives

$$\mathbf{r}(\tau) := D\psi(\tau) \circ \mathbf{s}(\tau), \qquad (3.28)$$

where $\psi(\tau)$ is convex, $\mathbf{s}(\tau)$ is measure preserving, and we drop the superscript *. Next put

$$\psi(\tau) = \psi_0 + \tau b(\tau), \ \mathbf{s}(\tau) = \mathbf{s_0} + \tau \mathbf{c}(\tau) \qquad (3.29)$$

into (3.28), differentiate with respect to τ, and set $\tau = 0$. We deduce

$$\mathbf{a} = Db + \mathbf{c}, \qquad (3.30)$$

for $b = b(0)$, $\mathbf{c} = \mathbf{c}(0)$. This is a Helmholtz decomposition of $\mathbf{a}$, since

$$1 = \det(D\mathbf{s}(\tau)) = 1 + \tau \text{div } \mathbf{c} + O(\tau^2) \qquad (3.31)$$

implies div $\mathbf{c} = 0$.   □

4 Regularity

4.1 Solving the Monge–Ampere equation

The mass allocation problem solved in §3 can be interpreted as providing a weak or generalized solution to the Monge–Ampere PDE mapping problem:

$$\begin{cases} \text{(a)} & f^-(D\phi(x))\det D^2\phi(x) = f^+(x) \quad (x \in X) \\ \text{(b)} & D\phi \text{ maps } X \text{ into } Y. \end{cases} \qquad (4.1)$$

Now and hereafter we omit the superscript *. Recall that we interpret (4.1)(a) to mean

$$\int_X h(D\phi(x))f^+(x)\,dx = \int_Y h(y)f^-(y)\,dy \qquad (4.2)$$

for each continuous function h.

There are, however, subtleties here. First of all, recall that since ϕ is convex, we can interpret $D^2\phi$ as a matrix of signed measures, which have absolutely continuous and singular parts with respect to Lebesgue measure:

$$d[D^2\phi] = [D^2\phi]_{ac}\,dx + d[D^2\phi]_s \qquad (4.3)$$

(see for instance [E-G2]). We say ϕ is a solution of

$$f^-(D\phi)\det(D^2\phi) = f^+ \qquad (4.4)$$

in the *sense of Alexandrov* if the identity (4.2) holds, and additionally

$$d[D^2\phi]_s = 0, \ \phi \text{ is strictly convex.} \qquad (4.5)$$

4.2   Examples

But, as noted by L. Caffarelli [C5], the solution constructed in §3 need *not* solve (4.4) in the Alexandrov sense. Consider first of all the situation that $n = 2$ and X is the unit disk $B(0,1)$. Then $\mathbf{s} = D\phi$, for

$$\phi(x_1, x_2) = |x_1| + \frac{1}{2}(x_1^2 + x_2^2),$$

optimally rearranges $\mu^+ = \chi_X dx$ into $\mu^- = \chi_Y \, dy$, where

$$
\begin{aligned}
Y \;=\; & \{(x_1, x_2) \mid 0 \le (x_1 - 1)^2 + x_2^2, \; x_1 \ge 1\} \\
& \cup \{(x_1, x_n) \mid 0 \le (x_1 + 1)^2 + x_2^2, \; x_1 \le -1\}.
\end{aligned}
$$

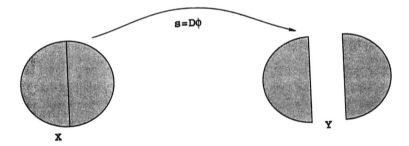

In this case $D^2\phi$ has a singular part concentrated along $\{x_1 = 0\}$. Caffarelli shows also by a perturbation argument that if we replace Y by a connected set Y_ε, as drawn, with $|Y_\varepsilon| = |Y| = |X|$, the optimal mapping $\mathbf{s}_\varepsilon$ still has a singularity.

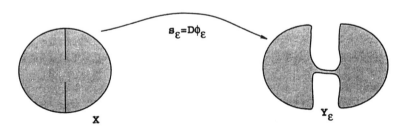

4.3 Interior regularity for convex targets

On the other hand Caffarelli also demonstrated that if the target Y is convex, the optimal mapping $s = D\phi$ is indeed regular. More precisely, assume that X, Y are bounded, connected, open sets in $\mathbb{R}^n$ and

$$f^+ : X \to (0, \infty), \ f^- : Y \to (0, \infty)$$

are bounded above and below, away from zero.

Theorem 4.1 *Assume Y is convex.*

(i) *Then an optimal mapping $s = D\phi$ solves the Monge–Ampere equation*

$$f^-(D\phi)\det(D^2\phi) = f^+$$

in the Alexandrov sense, and ϕ is strictly convex.

(ii)*In addition, $\phi \in C^{1,\alpha}_{\text{loc}}(X)$ for some $0 < \alpha < 1$.*

(iii) *If f^+, f^- are continuous, then*

$$\phi \in W^{2,p}_{\text{loc}}(X) \qquad \text{for each } 1 \leq p < \infty.$$

(iv) *Finally if $f^+ \in C^\beta(X)$, $f^- \in C^\beta(Y)$ for some $0 < \beta < 1$, then*

$$\phi \in C^{2,\alpha}_{\text{loc}}(X) \qquad \text{for each } 0 < \alpha < \beta.$$

The idea of the proof is to show that $\phi : \mathbb{R}^n \to \mathbb{R}$ is an Alexandrov solution of

$$C_2\chi_X \leq \det D^2\phi \leq C_1\chi_X \quad \text{in } \mathbb{R}^n$$

for constants $C_2 \geq C_1 > 0$, and to apply then the deep regularity theory developed in [C1], [C2], [C3].

4.4 Boundary regularity for convex domain and target

Theorem 4.2 *Assume both X, Y are convex.*

(i) *Then $\phi \in C^{1,\alpha}(\bar{X})$, $\psi \in C^{1,\alpha}(\bar{Y})$ for some $0 < \alpha < 1$.*

(ii) *If, in addition, $\partial X, \partial Y$ are smooth, then $\phi \in C^{2,\alpha}(\bar{X})$, $\psi \in C^{2,\alpha}(\bar{Y})$.*

These assertions have been proved independently by Urbas [U] and Caffarelli [C6], [C7], under slightly different hypotheses and using quite different techniques.

5 Application: Nonlinear interpolation

The next three sections discuss several new and interesting applications of the ideas set forth above in §2-4.

We begin with a clever procedure, due to R. McCann [MC1], which resolves a uniqueness problem in the calculus of variations. We consider a model for the *equilibrium configuration of an interacting gas*. Take

$$\mathcal{P} := \{\rho : \mathbb{R}^n \to \mathbb{R} \mid \rho \in L^1(\mathbb{R}^n), \ \rho \geq 0 \text{ a.e., } \int_{\mathbb{R}^n} \rho \, dx = 1\} \tag{5.1}$$

to be the collection of mass densities, and to each $\rho \in \mathcal{P}$ associate the *energy*

$$E(\rho) := \int_{\mathbb{R}^n} \rho^\gamma(x) \, dx + \int_{\mathbb{R}^n} \int_{\mathbb{R}^n} \rho(x) V(x-y)\rho(y) \, dxdy. \tag{5.2}$$

The first term

$$E_1(\rho) := \int_{\mathbb{R}^n} \rho^\gamma(x) \, dx$$

represents the *internal energy*, where $\gamma > 1$ is a constant. The second expression

$$E_2(\rho) := \int_{\mathbb{R}^n} \int_{\mathbb{R}^n} \rho(x) V(x-y)\rho(y) \, dxdy$$

corresponds to the *potential energy* owing to interaction, where $V : \mathbb{R}^n \to \mathbb{R}$ is nonnegative and strictly convex.

Suppose now $\rho_0, \rho_1 \in \mathcal{P}$ are both minimizers:

$$E(\rho_0) = E(\rho_1) = \min_{\rho \in \mathcal{P}} E(\rho). \tag{5.3}$$

A basic question concerns the uniqueness of these minimizers, up to translation invariance. Now the usual procedure would be to take the linear interpolation of ρ_0, ρ_1 and to ask if the mapping $t \mapsto E((1-t)\rho_1 + t\rho_0)$ is convex on $[0,1]$. This is however false in general for the example at hand. McCann instead employs the ideas of §2-3 to build a sort of "nonlinear interpolation" between ρ_0, ρ_1. His idea is to take the convex potential ϕ such that

$$\mathbf{s}_\#(\rho_0) = \rho_1, \tag{5.4}$$

where

$$\mathbf{s} = D\phi. \tag{5.5}$$

Define then

$$\rho_t := ((1-t)id + t\mathbf{s})_{\#}\rho_0 \quad (0 \le t \le 1). \tag{5.6}$$

This nonlinear interpolation between ρ_0 and ρ_1 in effect locates a realm of convexity for E.

Theorem 5.1 *The mapping* $t \mapsto E(\rho_t)$ *is convex on* $[0,1]$, *and is strictly convex unless* ρ_1 *is a translate of* ρ_0.

Proof. 1. We first compute

$$
\begin{aligned}
E_2(\rho_t) &= \int_{\mathbb{R}^n} \int_{\mathbb{R}^n} \rho_t(x) V(x-y) \rho_t(y) \, dx dy \tag{5.7} \\
&= \int_{\mathbb{R}^n} \int_{\mathbb{R}^n} \rho_0(x) V((1-t)(x-y) + t(\mathbf{s}(x) - \mathbf{s}(y))) \rho_0(y) \, dx dy.
\end{aligned}
$$

As V is convex, the mapping $t \mapsto E_2(\rho_t)$ is convex.
   2. Next, (5.6) implies

$$
\begin{aligned}
E_1(\rho_t) &= \int_{\mathbb{R}^n} \rho_t^\gamma(x) \, dx \\
&= \int_{\mathbb{R}^n} \left[\frac{\rho_0}{\det[(1-t)I + tD\mathbf{s}]} \right]^\gamma \det[(1-t)I + tD\mathbf{s}] \, dx. \tag{5.8}
\end{aligned}
$$

Now the matrix

$$A_t := (1-t)I + tD\mathbf{s} = (1-t)I + tD^2\phi$$

is positive definite for $0 \le t < 1$ and a.e. x. (In particular, where $d[D^2\phi]_s = 0$.) Now if A is a positive-definite, symmetric matrix, we have

$$(\det A)^{1/n} = \frac{1}{n} \inf_{\substack{B > 0 \\ \det B = 1}} (A:B),$$

and thus $A \mapsto (\det A)^{1/n}$ is concave for positive definite matrices. Consequently

$$\alpha(t) := (\det[(1-t)I + tD\mathbf{s}])^{1/n} \text{ is concave on } [0,1]. \tag{5.9}$$

Now set

$$\beta(t) := \alpha(t)^{n(1-\gamma)} \quad (0 \le t \le 1).$$

Then
$$\beta''(t) = n(1-\gamma)(n - n\gamma - 1)\alpha(t)^{n(1-\gamma)-2}(\alpha')^2$$
$$+ n(1-\gamma)\alpha(t)^{n(1-\gamma)-1}\alpha'' \geq 0,$$

since $\gamma > 1$. Thus

$$t \mapsto \beta(t) \text{ is convex on } [0,1]. \tag{5.10}$$

Return to (5.8), which we rewrite to read

$$E_1(\rho_t) = \int_{\mathbb{R}} \rho_0 \beta(t) \, dx.$$

It follows that $t \mapsto E_1(\rho_t)$ is convex.

3. To show a minimizer is unique up to translation, we go back to (5.7). As V is strictly convex, the mapping $t \mapsto E_2(\rho_t)$ is strictly convex, unless $x - y = \mathbf{s}(x) - \mathbf{s}(y)$ ρ_0 a.e. Now such strict convexity would violate the fact that ρ_0, ρ_1 are minimizers. Consequently

$$x - \mathbf{s}(x) \text{ is independent of } x, \qquad \rho_0 \text{ a.e.,}$$

and so ρ_1 is a translate of ρ_0. □

A related technique for understanding the equilibrium shape of crystals is presented in McCann [MC3]. See also Barthe [Ba] for the derivation of various inequalities using mass transport methods.

6 Application: Time-step minimization and nonlinear diffusion

An interesting emerging theme in much current research is the interplay between Monge–Kantorovich mass transform problems and partial differential equations involving time. A nice example, based upon Otto [O1] and Jordan–Kinderlehrer–Otto [J-K-O1], [J-K-O2], concerns a time-step minimization approximation to nonlinear diffusion equation.

Preparatory to describing this procedure, let us first define for the functions $f^+, f^- \in \mathcal{P}$ the *Wasserstein distance*

$$d(f^+, f^-)^2 := \inf \left\{ \frac{1}{2} \int_{\mathbb{R}^n} \int_{\mathbb{R}^n} |x - y|^2 \, d\mu(x,y) \right\}, \tag{6.1}$$

the infimum taken over all nonnegative Radon measure μ whose projections are $d\mu^+ = f^+ dx$, $d\mu^- = f^- dy$. In view of §1, $d(f^+, f^-)^2$ is least cost of the Monge–Kantorovich mass reallocation of μ^+ to μ^-, for $c(x,y) = \frac{1}{2}|x - y|^2$.

6.1 Discrete time approximation

We next initiate a *time stepping procedure*, by first taking a small step size $h > 0$ and an initial profile $g \in \mathcal{P}$. We set $u^0 = g$ and then inductively define $\{u^k\}_{k=1}^{\infty} \subset \mathcal{P}$ by taking $u^{k+1} \in \mathcal{P}$ to minimize the functional

$$N_k(v) := \frac{d(v, u^k)^2}{h} + \int_{\mathbb{R}^n} \beta(v)\, dx \qquad (6.2)$$

among all $v \in \mathcal{P}$. Here $\beta : \mathbb{R}^n \to \mathbb{R}$ is a given convex function, with superlinear growth. We suppose

$$\int_{\mathbb{R}^n} \beta(u^0)\, dx < \infty. \qquad (6.3)$$

Convexity and weak convergence arguments show that there exists $u^{k+1} \in \mathcal{P}$ satisfying

$$N_k(u^{k+1}) = \min_{v \in \mathcal{P}} N_k(v) \qquad (k = 0, \dots). \qquad (6.4)$$

We can envision (6.2), (6.4) as a discrete, dissipation evolution, in which at step $k+1$ the updated density u^{k+1} strikes a balance between minimizing the potential energy $\int_{\mathbb{R}^n} \beta(v)\, dx$ and the distance $\frac{d(v, u^k)^2}{2h}$ to the previous density at step k.

Taking $v = u^k$ on the right-hand side of (6.3), we have

$$\frac{d(u^{k+1}, u^k)^2}{2h} + \int_{\mathbb{R}^n} \beta(u^{k+1})\, dx \leq \int_{\mathbb{R}^n} \beta(u^k)\, dx;$$

and so

$$\frac{1}{2h} \sum_{k=1}^{\infty} d(u^{k+1}, u^k)^2, \; \max_{k \geq 1} \int_{\mathbb{R}^n} \beta(u^{k+1})\, dx \leq \int_{\mathbb{R}^n} \beta(u^0)\, dx < \infty. \qquad (6.5)$$

Next, define $u_h : [0, \infty) \to P$ by setting $t_k = hk$, $u_h(t_k) = u^k$, and taking u_h to be linear on each time interval $[t_k, t_{k+1}]$ $(k = 0, 1, \dots)$.

The basic question is this: what happens when the step size h goes to 0?

6.2 Euler–Lagrange equation

To gain some insight, we follow Otto [O1] to compute the first variation of the minimization principle (6.4). For this, let us define

$$\alpha(z) := \beta'(z)z - \beta(z) \qquad (z \in \mathbb{R}) \qquad (6.6)$$

and also write

$$d(u^{k+1}, u^k)^2 = \frac{1}{2} \int_{\mathbb{R}^n} \int_{\mathbb{R}^n} |x - y|^2 \, d\mu_{k+1}(x, y), \tag{6.7}$$

where

$$\text{proj}_x \mu_{k+1} = u^{k+1} dx, \quad \text{proj}_y \mu_{k+1} = u^k dy. \tag{6.8}$$

In other words, the measure μ_{k+1} solves the relaxed problem.

Lemma 6.1 *Let* $\boldsymbol{\xi} \in C_c^\infty(\mathbb{R}^n; \mathbb{R}^n)$ *be a smooth, compactly supported vector field. Then*

$$\int_{\mathbb{R}^n} \int_{\mathbb{R}^n} \frac{(x - y)}{h} \cdot \boldsymbol{\xi}(x) \, d\mu_{k+1}(x, y) = \int_{\mathbb{R}^n} \alpha(u^{k+1}) \text{div} \, \boldsymbol{\xi} \, dx. \tag{6.9}$$

Proof. 1. We employ $\boldsymbol{\xi}$ to generate a domain variation, as follows. First solve the ODE

$$\begin{cases} \dot{\boldsymbol{\Phi}} = \boldsymbol{\xi}(\boldsymbol{\Phi}) & \left(\cdot = \frac{d}{d\tau} \right) \\ \boldsymbol{\Phi}(0) = x, \end{cases} \tag{6.10}$$

and then write $\boldsymbol{\Phi} = \boldsymbol{\Phi}(\tau, x)$ to display the dependence of $\boldsymbol{\Phi}$ on the parameter τ and the initial point x. Then for each τ, the mapping $x \mapsto \boldsymbol{\Phi}(\tau, x)$ is a diffeomorphism. Thus for small τ, we can implicitly define

$$u_\tau : \mathbb{R}^n \to \mathbb{R}$$

by the formula

$$(\det D\boldsymbol{\Phi}(x, \tau)) u_\tau(\boldsymbol{\Phi}(\tau, x)) = u^{k+1}(x). \tag{6.11}$$

Clearly then

$$\int_{\mathbb{R}^n} u_\tau \, dx = \int_{\mathbb{R}^n} u^{k+1} \, dx = 1.$$

Thus $u_\tau \in \mathcal{P}$, and so

$$i(\tau) := N_k(u_\tau) \text{ has a minimum at } \tau = 0. \tag{6.12}$$

2. Now if $z = \boldsymbol{\Phi}(\tau, x)$, then

$$\int_{\mathbb{R}^n} \beta(u_\tau) dz = \int_{\mathbb{R}^n} \beta(u^{k+1}(\det D\boldsymbol{\Phi})^{-1}) \det D\boldsymbol{\Phi} \, dx.$$

Since $\frac{d}{dt}(\det D\Phi) = \operatorname{div} \boldsymbol{\xi}$ at $\tau = 0$, we compute

$$\frac{d}{dt} \int_{\mathbb{R}^n} \beta(u_\tau) dz\big|_{\tau=0} = -\int_{\mathbb{R}^n} \alpha(u^{k+1}) \operatorname{div} \boldsymbol{\xi}\, dx, \qquad (6.13)$$

owing to (6.6).

3. Next define $\mu_\tau \in \mathcal{M}_k$ so that

$$\frac{1}{2} \int_{\mathbb{R}^n} \int_{\mathbb{R}^n} |z - y|^2\, d\mu_\tau(z, y) = \frac{1}{2} \int_{\mathbb{R}^n} \int_{\mathbb{R}^n} |\Phi(\tau, x) - y|^2\, d\mu_{k+1}(x, y).$$

Then for $\tau > 0$:

$$\frac{1}{\tau}(d(u_\tau, u^k)^2 - d(u^{k+1}, u^k)^2) \leq \frac{1}{2\tau} \int_{\mathbb{R}^n} \int_{\mathbb{R}^n} (|\Phi(\tau, x) - y|^2 - |x - y|^2)\, d\mu_{k+1}(x, y).$$

Let $\tau \to 0^+$,

$$\frac{1}{h} \limsup_{\tau \to 0^+} \left[\frac{d(u_\tau, u^k)^2 - d(u^{k+1}, u^k)^2}{\tau} \right] \leq \int_{\mathbb{R}^n} \int_{\mathbb{R}^n} \frac{(x - y)}{h} \cdot \boldsymbol{\xi}\, d\mu_{k+1}.$$

Replacing $\boldsymbol{\xi}$ by $-\boldsymbol{\xi}$ and recalling (6.4), (6.12), we obtain (6.9). $\qquad \square$

6.3 Convergence

Let us suppose now that as $h \to 0$,

$$u_h \to u \text{ strongly in } L^1_{\text{loc}}(\mathbb{R}^n \times (0, \infty)). \qquad (6.14)$$

Theorem 6.1 *The function u is a weak solution of the nonlinear diffusion problem*

$$\begin{cases} u_t = \Delta\alpha(u) & \text{in } \mathbb{R}^n \times (0, \infty) \\ u = g & \text{on } \mathbb{R}^n \times \{t = 0\}. \end{cases} \qquad (6.15)$$

Notice from (6.6) that

$$\alpha'(z) = \beta''(z)z \geq 0 \quad (z \geq 0),$$

and so this nonlinear PDE is parabolic, corresponding to a nonlinear diffusion.

Proof. Let us fix $\zeta \in C_c^\infty(\mathbb{R}^n)$ and take $\boldsymbol{\xi} := D\zeta$. Then

$$\left| \int_{\mathbb{R}^n} u^{k+1}(x)\zeta(x)\, dx - \int_{\mathbb{R}^n} u^k(y)\zeta(y)\, dy - \int_{\mathbb{R}^n} \int_{\mathbb{R}^n} (x - y) \cdot D\zeta(x)\, d\mu_{k+1}(x, y) \right|$$

$$= \left| \int_{\mathbb{R}^n} \int_{\mathbb{R}^n} [\zeta(x) - \zeta(y) - (x - y) \cdot D\zeta(x)]\, d\mu_{k+1} \right|$$

$$\leq C \int_{\mathbb{R}^n} \int_{\mathbb{R}^n} |x - y|^2\, d\mu_{k+1} = Cd(u^{k+1}, u^k)^2$$

for some constant C. Consequently if $\phi \in C_c^\infty(\mathbb{R}^n \times [0, \infty))$, the Lemma
lets us estimate

$$\left| -\int_0^\infty \int_{\mathbb{R}^n} u_n \frac{(\phi(\cdot, t+h) - \phi(\cdot, t))}{h} \, dx dt - \int_h^\infty \int_{\mathbb{R}^n} \alpha(u_h) \Delta \phi \, dx dt - \frac{1}{h} \int_0^h \int_{\mathbb{R}^n} g\phi \, dx dt \right|$$

$$\leq C \sum_{k=1}^\infty d(u^{k+1}, u^k)^2 \leq Ch.$$

Owing then to (6.5),

$$\int_0^\infty \int_{\mathbb{R}^n} -u\phi_t - \alpha(u)\Delta\phi \, dx dt - \int_{\mathbb{R}^n} g\phi(\cdot, 0) \, dx = 0.$$

That this identity hold for all ϕ as above means u is a weak solution of
the nonlinear diffusion PDE.   □

   Remark. An interesting example is

$$\beta(z) = z \log z \quad (z > 0),$$

in which case the term

$$\int_{\mathbb{R}^n} \beta(v) dx = \int_{\mathbb{R}^n} v \log v \, dx$$

corresponds to *entropy*, and

$$\alpha(z) = \beta'(z)z - \beta(z) = z.$$

So the usual linear heat equation results from this time-stepping procedure. It is best, however, to continue to regard each u^k as a density,
which at each stage balances the Wasserstein distance from the previous density against the entropy production. As explained in Jordan–
Kinderlehrer–Otto [J-K-O1], we can therefore envision the individual
approximations u^k as something like "Gibbs states", in a sort of local
equilibrium.   □

7   Application: Semigeostrophic models in meteorology

In this section we explain the remarkable connections between Monge–
Kan-torovich theory and the so-called semigeostrophic equations from
meteorology, a model for large scale, stratified atmospheric flows with
front formation. The most accessible reference for mathematicians is
Cullen–Norbury–Purser [C-N-P] (but see also [C-P1],[C-P2],[C-P3],[C-S].)

7.1   The PDE in physical variables

In this system of PDE there are seven unknowns:

$$\begin{cases} \mathbf{v}_g & = & (v_g^1, v_g^2, 0) = \text{ geostrophic wind velocity} \\ \mathbf{v}_a & = & (v_a^1, v_a^2, v_a^3) = \text{ ageostrophic wind velocity} \\ p & = & \text{pressure} \\ \theta & = & \text{potential temperature,} \end{cases} \tag{7.1}$$

defined in $X \times [0, \infty)$, X denoting a fixed region in $\mathbb{R}^3$. We also set

$$\mathbf{v} := \mathbf{v}_g + \mathbf{v}_a = \text{ total wind velocity.}$$

Write $x = (x_1, x_2, x_3) \in \mathbb{R}^3$, $D_x = \left(\frac{\partial}{\partial x_1}, \frac{\partial}{\partial x_2}, \frac{\partial}{\partial x_3} \right)$, and define the *convective derivative*

$$\frac{D_x}{Dt} := \frac{\partial}{\partial t} + \mathbf{v} \cdot D_x = \frac{\partial}{\partial t} + v^1 \frac{\partial}{\partial x_1} + v^2 \frac{\partial}{\partial x_2} + v^3 \frac{\partial}{\partial x_3}. \tag{7.2}$$

The *semigeostrophic equations* are then these seven equations:

$$\begin{cases} \text{(i)} & \dfrac{D_x v_g^1}{Dt} - v_a^2 = 0, \quad \dfrac{D_x v_g^2}{Dt} + v_a^1 = 0 \\ \text{(ii)} & \dfrac{D_x \theta}{Dt} = 0 \\ \text{(iii)} & \text{div}(\mathbf{v}) = 0 \\ \text{(iv)} & D_x p = (v_g^2, -v_g^1, \theta), \end{cases} \tag{7.3}$$

where we have set all physical parameters to 1. The equations (i) represent a simplification of the full Euler equations in regimes where centrifugal forces dominate, and (ii) is the passive transport of the potential θ with the flow. The equality (iii) of course means incompressibility. The first two components of (iv) embody the definition of the geostrophic wind $\mathbf{v}_g$ and the last is the definition of θ. Observe that v_a^1, v_a^2 are defined by (i) and that v_a^3 enters the system of PDE only through (iii).

We couple (7.3) with appropriate initial conditions and the boundary condition

$$\mathbf{v} \cdot \boldsymbol{\nu} = 0 \qquad \text{on } \partial U, \tag{7.4}$$

$\boldsymbol{\nu}$ being the outward unit normal along ∂U.

7.2   The PDE in dual variables

The system (7.3) is quite complicated, but its structure clarifies under an appropriate change of variable. We hereafter introduce new functions

$$\mathbf{y} = (y^1, y^2, y^3) := (x_1 + v_g^2, x_2 - v_g^1, \theta) \qquad (7.5)$$

and

$$s := \det(D_x \mathbf{y}). \qquad (7.6)$$

Then (7.3)(i),(ii) transform to read

$$\frac{D_x \mathbf{y}}{Dt} = \mathbf{v}_g, \qquad (7.7)$$

and

$$\frac{D_x s}{Dt} = 0. \qquad (7.8)$$

We have in mind next to change to new independent variables $y = (y_1, y_2, y_3)$. To do so introduce

$$\phi := \frac{1}{2}(x_1^2 + x_2^2) + p. \qquad (7.9)$$

Then

$$\mathbf{y} = D_x \phi. \qquad (7.10)$$

Consequently (7.6) says

$$s = \det(D_x^2 \phi). \qquad (7.11)$$

We assume ϕ is convex, and write ψ for its convex dual, $\psi = \psi(y)$. Then

$$\mathbf{x} = D_y \psi, \qquad (7.12)$$

for $\mathbf{x} = (x^1, x^2, x^3)$. If we write

$$r := \det(D_y \mathbf{x}) = \det(D_y^2 \psi), \qquad (7.13)$$

then using (7.7), (7.8) we can compute

$$\frac{D_y r}{Dt} = 0, \qquad (7.14)$$

where

$$\frac{D_y}{Dt} := \frac{\partial}{\partial t} + \mathbf{v}_g \cdot D_y = \frac{\partial}{\partial t} + v_g^1 \frac{\partial}{\partial y_1} + v_g^2 \frac{\partial}{\partial y_2}. \qquad (7.15)$$

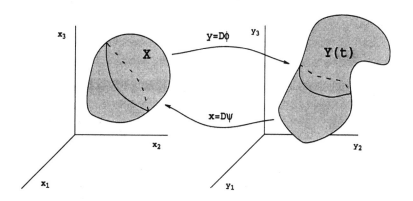

Now let $Y(t)$ denote the range of $D_x\phi : X \to \mathbb{R}^3$, at time t. Then we can in summary rewrite (7.13), (7.14) to read

$$\begin{cases} \text{(i)} & r_t + \text{div}(r\mathbf{w}) = 0 \\ \text{(ii)} & \mathbf{w} = (\psi_{y_2} - y_2, -(\psi_{y_1} - y_1), 0) \quad \text{in } Y(t) \times [0, \infty) \\ \text{(iii)} & \det(D^2\psi) = r, \end{cases} \qquad (7.16)$$

where we have set $\mathbf{w} := \mathbf{v}_g$. The additional requirement is

$$D\psi \in X. \qquad (7.17)$$

The system (7.16) is a sort of time-dependent Monge–Ampere equation involving the moving free boundary $\partial Y(t)$. Perhaps more interestingly, (7.16) is an obvious variant of the vorticity formulation of the two-dimensional Euler equations

$$\begin{cases} \text{(i)} & \omega_t + \text{div}(\omega\mathbf{v}) = 0 \\ \text{(ii)} & \mathbf{v} = (-\psi_{x_2}, \psi_{x_1}) \quad \text{in } \mathbb{R}^2 \times [0, \infty). \\ \text{(iii)} & \Delta\psi = \omega, \end{cases} \qquad (7.18)$$

where ψ is the stream function, and ω the scalar vorticity. Roughly speaking, the system (7.18) is a linearization of (7.16).

F. Otto [O2] and Benamou–Brenier [Be-B1] have shown the existence of a weak solution of (7.16), making use of the pair (ϕ, ψ) in the sense of the Monge–Kantorovich theory discussed above. One interpretation is that while r, $\mathbf{w}$ evolve on an "order-one" time scale, there is an optimal Monge–Kantorovich rearrangement of air parcels on a "fast" time scale. See also Brenier [B4].

7.3 Frontogenesis

We can informally interpret the dynamics (7.16) as supporting the onset of "fronts", i.e. surfaces of discontinuity in the velocity and temperature.

To understand this, suppose $X, Y(0)$ are uniformly convex, and for simplicity take $r \equiv 1$. Then (7.16)(i) is trivial and we can understand (7.16)(ii), (iii) as a law of evolution of $\partial Y(t)$. Suppose for heuristic purposes that $\partial Y(t)$ remains smooth. Then owing to the regularity theory (§4) $\psi(\cdot, t)$ will be smooth for each $t \geq 0$. But since $Y(t)$ is changing shape, it is presumably possible that $Y(t)$ is no longer convex at some sufficiently large time t, as illustrated.

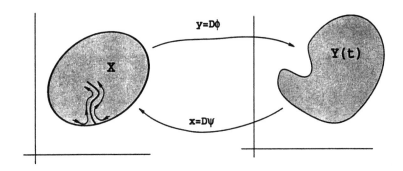

Then the regularity theory from §4 no longer applies to $\phi(\cdot, t)$. We may consequently expect that although $\psi(\cdot, t)$ remains smooth, $\phi(\cdot, t)$ will not.

Far from being a defect, the advent of singularities of $\phi(\cdot, t)$ in the physical variables $x = (x_1, x_2, x_3)$ is a definite advantage in the model. As discussed in Cullen [C], the meteorologists wish to model how fronts arise in large scale weather patterns. Tracking these fronts, that is, thin regions across which there are large variations in wind and temperature, is a central goal, and the semigeostrophic equations provide a plausible mechanism for frontogenesis. An interesting physical rationale occurs in [C], where the author points out that such discontinuities are *contact discontinuities* in the sense of fluid mechanics: this means that the air parcels move parallel to, and not across, them. (See the simplified, two-dimensional illustration.) Obviously in regions of rapid temperature and wind change the approximations which transform the full Euler equations into the semigeostrophic PDE are very suspect. But, since the proposed fronts are contact discontinuities, most of the air mass stays away from such regions, and so the various approximations should overall be pretty good.

There are extremely interesting mathematical problems here, which have only in small part been studied. There are likewise many issues concerning computing these flows: see Benamou [Be1],[Be2] and Benamou–Brenier [Be-B2] for this.

Part II: Cost = Distance

8 Heuristics

For the rest of this paper we turn attention to the nonuniformly convex cost density

$$c(x,y) = |x - y| \quad (x, y \in \mathbb{R}^n). \tag{8.1}$$

The task now is to find an optimal mass transfer plan $\mathbf{s}^*$ solving Monge's original problem (1.5), where

$$I[\mathbf{s}] := \int_{\mathbb{R}^n} |x - \mathbf{s}(x)| \, d\mu^+(x) \tag{8.2}$$

for $\mathbf{s} \in \mathcal{A}$, the admissible class of measurable, one-to-one maps of $\mathbb{R}^n$ which push μ^+ to μ^-.

It will turn out that the nonuniform convexity of the cost density (8.1) defeats any simple attempt to modify the techniques described before in §2–7. We will therefore first of all need some new insights, to help us sort out the structure of optimal mass allocations.

8.1 Geometry of optimal transport

As in §2 we start with heuristics, our intention being to read off useful information from a "twist variation".

So assume that $\mathbf{s}^* \in \mathcal{A}$ minimizes (8.2). Fix then $m \geq 2$, select any points $\{x_k\}_{k=1}^m \subset X = \mathrm{spt}\,(\mu^+)$, and small balls $E_k := B(x_k, r_k)$, the radii selected so that $\mu^+(E_1) = \cdots = \mu^+(E_m) = \varepsilon$. Next set $y_k = \mathbf{s}^*(x_k)$, $F_k := \mathbf{s}^*(E_k)$, and finally fix an integer $l \geq 1$. Similarly now to the construction in §2, we build $\mathbf{s} \in \mathcal{A}$ so that

$$
\begin{cases}
\mathbf{s}(x_k) = y_{k+l}, \ \mathbf{s}(E_k) = F_{k+l} \\
\mathbf{s} \equiv \mathbf{s}^* \text{ on } X - \bigcup_{k=1}^m E_k,
\end{cases}
$$

where we compute the subscripts mod m. As in §2 we deduce from the inequality $I[\mathbf{s}^*] \leq I[\mathbf{s}]$ that

$$
\sum_{k=1}^m |x_k - y_k| \leq \sum_{k=1}^m |x_k - y_{k+l}|. \tag{8.3}
$$

We as follows draw a geometric deduction from (8.3). Take any closed curve C lying in $X = \mathrm{spt}\,(\mu^+)$, with constant speed parameterization $\{\mathbf{r}(t) \mid 0 \leq t \leq 1\}$, $\mathbf{r}(0) = \mathbf{r}(1)$. Let $x_k := \mathbf{r}\left(\frac{k}{m}\right)$ be m equally spaced points along C.

Fix $\tau > 0$. Then using (8.3) and letting $m \to \infty$, $\frac{l}{m} \to \tau$, we deduce:

$$
\int_0^1 |\mathbf{r}(t) - \mathbf{s}^*(\mathbf{r}(t))|dt \ \leq \ \int_0^1 |\mathbf{r}(t) - \mathbf{s}^*(\mathbf{r}(t+\tau))|dt
$$

$$
= \ \int_0^1 |\mathbf{r}(t-\tau) - \mathbf{s}^*(\mathbf{r}(t))|dt =: i(\tau).
$$

Hence $i(\cdot)$ has a minimum at $\tau = 0$, and so

$$
0 = i'(0) = -\int_0^1 \frac{[\mathbf{r}(t) - \mathbf{s}^*(\mathbf{r}(t))] \cdot \mathbf{r}'(t)}{|\mathbf{r}(t) - \mathbf{s}^*(\mathbf{r}(t))|} dt.
$$

This is to say,

$$
\int_C \boldsymbol{\xi} \cdot d\mathbf{r} = 0, \tag{8.4}
$$

for the vector field

$$
\boldsymbol{\xi}(x) := \frac{\mathbf{s}^*(x) - x}{|\mathbf{s}^*(x) - x|}.
$$

As (8.4) holds for all closed curves C, we conclude that $\boldsymbol{\xi}$ is a gradient:

$$
\frac{\mathbf{s}^*(x) - x}{|\mathbf{s}^*(x) - x|} = -Du^*(x) \tag{8.5}
$$

for some scalar mapping $u^* : \mathbb{R}^n \to \mathbb{R}$, which solves the *eikonal equation*

$$|Du^*| = 1 \qquad (8.6)$$

on $X = \text{spt}(\mu^+)$. In other words, the direction $\mathbf{s}^*$ maps each point x is given by the gradient of a potential:

$$\mathbf{s}^*(x) = x - d^*(x)Du^*(x).$$

Note very carefully however that this deduction provides absolutely no information about the distance $d^*(x) := |\mathbf{s}^*(x) - x|$.

Remark. The foregoing insights were first discovered by Monge, based upon completely different, geometric arguments concerning developable surfaces. See Monge [M], Dupin [D]. Monge's and his students' investigations of this problem lead to many of their foundational discoveries in differential geometry: see for instance Struik [St]. □

8.2 Lagrange multipliers

It is interesting to see also a formal, analytic derivation of (8.5), (8.6). The following computations are essentially those of Appell [A], from the turn of the century.

We assume for this that $d\mu^+ = f^+dx$, $d\mu^- = f^-dy$ for smooth densities $f^\pm$. The *augmented work functional* is

$$\tilde{I}[\mathbf{s}] := \int_{\mathbb{R}^r} |x - \mathbf{s}(x)|f^+(x) + \lambda(x)[f^-(\mathbf{s}(x))\det(D\mathbf{s}(x)) - f^+(x)]\,dx,$$

where λ is the *Lagrange multiplier* for the constraint $f^+ = f^-(\mathbf{s})\det(D\mathbf{s})$. The first variation is

$$(\lambda f^-(\mathbf{s}^*)(\text{cof}D\mathbf{s}^*)_i^k)_{x_i} = \frac{s^{*k} - x_k}{|\mathbf{s}^* - x|}f^+(x) + \lambda f_{y_k}^-(\mathbf{s}^*)\det D\mathbf{s}^*.$$

Simplifying as in §2, we deduce

$$\lambda x_j = \frac{s^{*k} - x_k}{|\mathbf{s}^* - x|}s_{x_j}^{*k} \quad (j = 1, \dots, n).$$

Next define u^* by

$$u^*(\mathbf{s}^*(x)) = -\lambda(x).$$

Then

$$u_{y_k}^* s_{x_j}^{*k} = -\lambda x_j = \frac{s^{*k} - x_k}{|\mathbf{s}^* - x|}s_{x_j}^{*k}.$$

As Ds^* is invertible, we see

$$Du^*(s^*(x)) = -\frac{s^* - x}{|s^* - x|}.$$

But $Du^*(\mathbf{s}^*(x)) = Du^*(x)$, and so (8.5), (8.6) again follow.

Remark. It is instructive to compare and contrast the heuristics of this section with those in §2. The argument based upon the "twist variation" in §8.1 is more general than that using cyclic monotonicity in §2.1, and adapts without trouble to a general cost density $c(x,y)$. The conclusion is then that

$$D_x c(x, \mathbf{s}^*(x)) = Du^*(x) \text{ for some scalar potential function } u^*.$$
$$(8.7)$$

Now *if* we can invert (8.7), to solve for $\mathbf{s}^*$ in terms of Du^*, we thereby derive a structural formula for an optimal mapping. Much of the interest in (8.1) is precisely that this inversion is not possible if c is not uniformly convex. $\square$

9 Optimal mass transport

9.1 Solution of dual problem

Our intention now is to transform the foregoing heuristic calculations into a proof of the existence of an optimal mapping $\mathbf{s}^*$, where we expect

$$\mathbf{s}^*(x) = x - d^*(x)Du^*(x) \qquad (9.1)$$

for some potential u^*, with $|Du^*(x)| = 1$ and so $d^*(x) = |\mathbf{s}^*(x) - x|$.
  We turn to the case

$$d\mu^+ = f^+ dx, \ d\mu^- = f^- dy \qquad (9.2)$$

where $f^\pm$ are bounded, nonnegative functions with compact support, satisfying the *mass balance* compatibility condition

$$\int_X f^+(x)\, dx = \int_Y f^-(y)\, dy \qquad (9.3)$$

where $X = \text{spt}(f^+)$, $Y = \text{spt}(f^-)$.
  We further remember the dual problem (§1), which asks us to find (u^*, v^*) to *maximize*

$$K[u,v] := \int_X u(x)f^+(x)\, dx + \int_Y v(y)f^-(y)\, dy \qquad (9.4)$$

subject to

$$u(x) + v(y) \leq |x - y| \quad (x \in X, \ y \in Y). \tag{9.5}$$

Lemma 9.1 (i) *There exist* (u^*, v^*) *solving this maximization problem.*
(ii) *Furthermore, we can take*

$$v^* = -u^*, \tag{9.6}$$

where $u^* : \mathbb{R}^n \to \mathbb{R}$ *is Lipschitz continuous, with*

$$|u^*(x) - u^*(y)| \leq |x - y| \quad (x, y \in \mathbb{R}^n). \tag{9.7}$$

Proof. 1. If u, v satisfy (9.5), then

$$u(x) \leq \min_{y \in Y}(|x - y| - v(y)) =: \hat{u}(x) \tag{9.8}$$

and

$$\hat{u}(x) + v(y) \leq |x - y| \quad (x \in X, \ y \in Y).$$

Therefore

$$v(y) \leq \min_{x \in X}(|x - y| - \hat{u}(x)) =: \hat{v}(y) \tag{9.9}$$

and

$$\hat{u}(x) + \hat{v}(y) \leq |x - y| \quad (x \in X, \ y \in Y). \tag{9.10}$$

Furthermore, since $\hat{v} \geq v$, (9.8) implies

$$\hat{u}(x) \geq \min_{y \in Y}(|x - y| - \hat{v}(y));$$

and so (9.10) implies

$$\hat{u}(x) = \min_{y \in Y}(|x - y| - \hat{v}(y)). \tag{9.11}$$

Since $f^{\pm} \geq 0$ and $\hat{u} \geq u$, $\hat{v} \geq v$ we see that $K[u, v] \leq K[\hat{u}, \hat{v}]$.

2. Thus in seeking to maximize K we may restrict attention to "dual" pairs $(\hat{u}, \hat{v})$, as above. But then

$$\hat{u} + \hat{v} = 0 \text{ on } X \cap Y. \tag{9.12}$$

To see this, take $z \in X \cap Y$ (if $X \cap Y \neq \emptyset$), and suppose

$$\hat{u}(z) + \hat{v}(z) < 0. \tag{9.13}$$

Take $x \in X$, $y \in Y$ so that

$$\begin{cases} \hat{u}(z) = |z - y| - \hat{v}(y) \\ \hat{v}(z) = |x - z| - \hat{u}(x). \end{cases}$$

Rearrange and add:

$$\begin{aligned} |x - z| + |z - y| &= \hat{u}(x) + \hat{v}(y) + \hat{u}(z) + \hat{v}(z) \\ &< \hat{u}(x) + \hat{v}(y) \text{ by (9.13)} \\ &\leq |x - y| \text{ by (9.10)}. \end{aligned}$$

This contradiction shows $\hat{u} + \hat{v} \geq 0$ on $X \cap Y$. As (9.10) clearly implies $\hat{u} + \hat{v} \leq 0$ on $X \cap Y$, assertion (9.12) follows.

3. In view of (9.12) let us extend the definition of $\hat{u}$ by setting

$$\hat{u} := -\hat{v} \qquad \text{on } Y.$$

Then (9.10) reads

$$\hat{u}(x) - \hat{u}(y) \leq |x - y| \qquad (x \in X, \; y \in Y).$$

Finally extend $\hat{u}$ to all of $\mathbb{R}^n$, keeping the Lipschitz constant:

$$|\hat{u}(x) - \hat{u}(y)| \leq |x - y| \qquad (x, y \in \mathbb{R}^n).$$

Our problem is thus to maximize

$$K[u] := \int_{\mathbb{R}^n} u(f^+ - f^-) \, dz, \qquad (9.14)$$

subject to the Lipschitz constraint

$$|u(x) - u(y)| \leq |x - y| \qquad (x, y \in \mathbb{R}^n). \qquad (9.15)$$

This problem clearly admits a solution u^*, and we define $v^* := -u^*$ to obtain a pair solving the original dual problem. $\qquad \square$

Note that the Lipschitz condition implies that Du^* exists a.e. in $\mathbb{R}^n$, with

$$|Du^*| \leq 1 \quad \text{a.e. }.$$

We further expect

$$|Du^*| = 1 \quad \text{a.e. on } X \cup Y.$$

9.2   Existence of optimal mass transfer plan

We wish next to employ u^* to build an optimal mass allocation mapping $\mathbf{s}^*$. This is not so easy as in the uniformly convex case that $c(x, y) = \frac{1}{2}|x - y|^2$, discussed in §3. The central problem is that although we expect $\mathbf{s}^*$ to have the structure (9.1) there is still an unknown, namely the *distance* $d^*(x) = |\mathbf{s}^*(x) - x|$ that the point x should move: $Du^*(x)$ tells us only the *direction*.

This problem was solved in important work by Sudakov [Su], using rather subtle measure theoretic techniques (cf. Rohlin [Ro]).

In keeping with the overall theme of this paper, we will here discuss instead an alternative, differential-equations-based procedure, from [E-G1].

To repeat, the basic issue is that we must somehow extract the missing information about the distance $d^*(x) = |\mathbf{s}^*(x) - x|$ from the variational problem (9.14), (9.15). It is convenient in doing so to introduce some standard notion from convex analysis. Let us set

$$\mathbb{K} := \{ u \in L^2(\mathbb{R}^n) \mid |Du| \leq 1 \text{ a.e.} \} \qquad (9.16)$$

and

$$I_\infty[u] := \begin{cases} 0 & \text{if } u \in \mathbb{K} \\ +\infty & \text{otherwise.} \end{cases} \qquad (9.17)$$

Then u^* minimizes $K[\cdot]$ over $\mathbb{K}$. The corresponding Euler Lagrange equation is

$$f^+ - f^- \in \partial I_\infty[u^*]; \qquad (9.18)$$

that is,

$$I_\infty[v] \geq I_\infty[u^*] + (f^+ - f^-, v - u^*)_{L^2}$$

for all $v \in L^2(\mathbb{R}^n)$.

We need firstly to convert (9.18) into more concrete form:

Lemma 9.2 *Assume additionally that $f^\pm$ are Lipschitz continuous.*
  (i) *Then there exists a nonnegative L^∞ function a such that*

$$- \operatorname{div}(a Du^*) = f^+ - f^- \quad \text{in } \mathbb{R}^n. \qquad (9.19)$$

  (ii) *Furthermore*

$$|Du| = 1 \quad \text{a.e. on the set } \{a > 0\}.$$

We call a the *transport density*. The PDE (9.19) looks linear, but is not: *the function a is the Lagrange multiplier corresponding to the constraint* $|Du^*| \leq 1$, and is a highly nonlinear and nonlocal function of u^*. On the other hand once u^* is known, (9.19) can be thought of as a linear, first-order PDE for a.

Outline of Proof. Take $n + 1 \leq p < \infty$. We approximate by the quasilinear PDE

$$- \operatorname{div}(|Du_p|^{p-2} Du_p) = f^+ - f^-, \qquad (9.20)$$

which corresponds to the problem of maximizing

$$K_p[u] := \int_{\mathbb{R}^n} u(f^+ - f^-) - \frac{1}{p} |Du|^p \, dz.$$

A maximum principle argument shows that

$$\sup_p |u_p|, |Du_p|, |Du_p|^p \leq C < \infty$$

for some constant C. (Cf. Bhattacharya–DiBenedetto–Manfredi [B-B-M].)

It follows that there exists a sequence $p_k \to \infty$ such that

$$\begin{cases} u_{p_k} \to u^* & \text{locally uniformity} \\ Du_{p_k} \to Du^* & \text{boundedly, a.e.} \\ |Du_{p_k}|^{p-2} \rightharpoonup a & \text{weakly } * \text{ in } L^\infty. \end{cases}$$

Then passing to limits in (9.20) we obtain (9.19). □

As noted above, a is the Lagrange multiplier from the constraint $|Du^*| \leq 1$. It turns out furthermore that a in fact "contains" the missing information as to the distance $d^*(x)$.

The recipe is to build $\mathbf{s}^*$ by solving a flow problem involving Du^*, a, etc. So fix $x \in X$ and consider the ODE

$$\begin{cases} \dot{\mathbf{z}}(t) = \mathbf{b}(\mathbf{z}(t), t) & (0 \leq t \leq 1) \\ \mathbf{z}(0) = x, \end{cases} \qquad (9.21)$$

for the time-varying vector field

$$\mathbf{b}(z, t) := \frac{-a(z) Du^*(z)}{t f^-(z) + (1-t) f^+(z)}. \qquad (9.22)$$

Ignore for the moment that a, Du^* are not smooth (or even continuous in general) and that we may be dividing by zero in (9.22). Proceeding formally then, let us write

$$\mathbf{s}^*(x) = \mathbf{z}(1), \qquad (9.23)$$

the time-one map of the flow.

Theorem 9.1 *Define* s* *by* (9.23). *Then*
(i) $s^* : X \to Y$ *is essentially one-to-one and onto.*
(ii) $\int_X h(s^*(x)) \, d\mu^+(x) = \int_Y h(y) \, d\mu^-(y)$ *for each* $h \in C(Y)$.
(iii) *Lastly,*

$$\int_X |x - s^*(x)| \, d\mu^+(x) \le \int_X |\mathbf{x} - \mathbf{s}(x)| \, d\mu^+(x)$$

for all $\mathbf{s} : X \to Y$ *such that* $\mathbf{s}_\#(\mu^+) = \mu^-$.

Outline of Proof. 1. Write

$$\mathbf{s}(t, x) := z(t) \quad (0 \le t \le 1),$$

$$J(z, t) := \det D\mathbf{s}(z, t).$$

Then $J_t = (\text{div } \mathbf{b})J$; and so, following Dacorogna–Moser [D-M], we may compute

$$
\begin{aligned}
\frac{\partial}{\partial t}&[(tf^-(\mathbf{s}(z,t)) + (1-t)f^+(\mathbf{s}(z,t)))J(z,t)] \\
&= (f^- - f^+)J + (tDf^- \cdot \mathbf{s}_t + (1-t)Df^+ \cdot \mathbf{s}_t)J \\
&\quad + (tf^- + (1-t)f^+)J_t \\
&= [(f^- - f^+) + (tDf^- \cdot \mathbf{b} + (1-t)Df^+ \cdot \mathbf{b}) \\
&\quad + (tf^- + (1-t)f^+)\text{div } \mathbf{b}]J.
\end{aligned}
\tag{9.24}
$$

But in view of (9.19), (9.22):

$$
\begin{aligned}
\text{div } \mathbf{b} &= \frac{f^+ - f^-}{tf^- + (1-t)f^+} + \frac{(tDf^- + (1-t)Df^+) \cdot (aDu^*)}{(tf^- + (1-t)f^+)^2} \\
&= \frac{f^+ - f^- - (tDf^- + (1-t)Df^+) \cdot \mathbf{b}}{tf^- + (1-t)f^+}.
\end{aligned}
$$

Consequently the last term in (9.24) is zero, and hence

$$f^-(\mathbf{s}^*)\det D\mathbf{s}^* = f^+.$$

This confirms assertion (ii).
2. We next verify s* is optimal. For this, take s to be any admissible mapping.

Then

$$\int_X |x - \mathbf{s}^*(x)|\, d\mu^+(x) = \int_X [u^*(x) - u^*(\mathbf{s}^*(x))] f^+(x)\, dx$$
$$= \int_X u^* f^+ - \int_Y u^* f^-\, dy$$
$$= \sup_{|Dw|\le 1} \int_{\mathbb{R}^n} w(f^+ - f^-)\, dz$$
$$= \min_{\mu\in\mathcal{M}} \iint |x - y|\, d\mu(x,y)$$
$$\le \int_X |x - \mathbf{s}(x)| f^+(x)\, dx.$$

The last equality holds owing to the duality of the primal and dual problems: cf. (13.1) in the appendix. □

This "Theorem" should of course really be in quotes, since the "proof" just outlined is purely formal. In reality neither a nor Du^* nor $f^\pm$ are smooth enough to justify these computations. A careful proof is available in [E-G1]: the full details are extremely complicated, involving a very careful smoothing of u^*. The proof requires as well the additional conditions that $\partial X, \partial Y$ are nice and $X \cap Y = \emptyset$, although these requirements are presumably not really necessary.

Remark. A nice paper by Cellina and Perrotta [C-P] discusses somewhat related issues. Jensen [Je] considers the subtle problem of what happens to the approximations u_p within the regions $\{f^+ \equiv f^- \equiv 0\}$, in the limit $p \to \infty$.

More on the connections with optimal flow problems may be found in Iri [I] and Strang [S1], [S2]. □

9.3 Detailed mass balance, transport density

For reference later we record here some properties of the optimal potential u^* and transport density a, proved in [E-G1].

First a.e. point $x \in X$ lies in a unique maximal line segment R_x along which u^* decreases linearly at rate one. We call R_x the *transport ray* through x. The idea is that we move the point x "downhill" along R_x to $y = \mathbf{s}^*(x)$, and the ODE (9.21), (9.22) tells us how far to go.

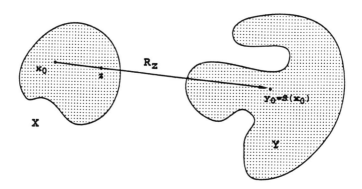

Next we note that these transport rays subdivide X and Y into subregions of equal μ^+, μ^- measure.

Lemma 9.3 *Let $E \subset \mathbb{R}^n$ be a measurable set with the property that for each point $x \in E$, the full transport ray R_x also lies in E. Then*

$$\int_{X \cap E} f^+ \, dx = \int_{Y \cap E} f^- \, dy. \qquad (9.25)$$

We call (9.25) the *detailed mass balance* relation. It asserts that the line segments along which u^* changes linearly with slope one naturally partition X and Y into subregions of equal masses. This must be so if our transport scheme (9.23) is to work.

Outline of Proof. We provide an interesting, but purely heuristic, derivation of a somewhat weaker statement. The trick is to employ a Hamilton–Jacobi PDE to generate a variation for the maximization principle (9.14), (9.15).

For this, take $H : \mathbb{R}^n \to \mathbb{R}$ to be any smooth Hamiltonian. We solve then the initial-value problem:

$$\begin{cases} w_t + H(Dw) = 0 & \text{in } \mathbb{R}^n \times \{t > 0\} \\ \qquad\quad w = u^* & \text{on } \mathbb{R}^n \times \{t = 0\}. \end{cases} \qquad (9.26)$$

Now the mapping $t \to w(\cdot, t)$ is a contraction in the sup-norm, and therefore for each time $t > 0$:

$$\|Dw(\cdot, t)\|_{L^\infty} \leq \|Du^*(\cdot, t)\|_{L^\infty} \leq 1.$$

Hence $w(\cdot, t)$ is a valid competitor in the variational principle. Since u^* solves (9.14), (9.15) and $w(\cdot, 0) = u^*$, it follows that

$$i(t) \leq i(0) \quad (t \geq 0),$$

where

$$i(t) := \int_{\mathbb{R}^n} w(\cdot, t)(f^+ - f^-) dz.$$

Hence $i'(0) \leq 0$. In view of (9.26) therefore,

$$\int_{\mathbb{R}^n} H(Du^*)(f^+ - f^-) dz = -i'(0) \geq 0.$$

Replacing H by $-H$, we conclude that

$$\int_{\mathbb{R}^n} H(Du^*)(f^+ - f^-) dz = 0 \qquad (9.27)$$

for *all* smooth $H : \mathbb{R}^n \to \mathbb{R}$.

Taking H to approximate χ_A, where $A \subset S^{n-1}$, we deduce from (9.27) that the detailed mass balance holds for the particular set $E := \{z \mid Du^*(z) \in A\}$.   □

This proof is not really rigorous as u^*, w are not smooth enough to justify the stated computations. See e.g. [E-G1] for a careful proof.

It is also useful to understand how smooth the transport density is along the transport rays, and we also need to check that the ODE flow (9.21) does not "overshoot" the endpoints.

Lemma 9.4 (i) *For a.e.* x, *the transport density* a, *restricted to* R_x, *is locally Lipschitz continuous.*

(ii) *Furthermore,*

$$\lim_{\substack{z \to a_0, b_0 \\ z \in R_x}} a(z) = 0,$$

where a_0, b_0 *are the endpoints of* R_x.

The first calculations in the literature related to assertion (ii) seem to be those of Janfalk [J].

10   Application: Shape optimization

The ensuing three sections describe some applications extending the ideas set forth in §8,9.

As a first application, we discuss the following shape optimization problem. Suppose we are given two nonnegative Radon measures $\mu^\pm$ on $\mathbb{R}^n$, with $\mu^+(\mathbb{R}^n) = \mu^-(\mathbb{R}^n)$. Think of μ^+ as giving the density of a given electric charge. We interpret $\mathbb{R}^n$ as an insulating medium, into which we place a fixed amount of some conducting material, whose conductivity

(= inverse resistivity) is described by a nonnegative measure σ. We imagine then the resulting steady current flow within $\mathbb{R}^n$ from μ^+ to μ^-, and ask if we can optimize the placement of the conducting material so as to *minimize the heating* induced by the flow.

To be more precise, consider the admissible class

$$\mathcal{S} = \{\text{nonnegative Radon measures } \sigma \text{ on } \mathbb{R}^n \mid \sigma(\mathbb{R}^n) = 1\}$$

and think of each $\sigma \in \mathcal{S}$ as describing how to arrange a unit quantity of conducting material within $\mathbb{R}^n$. Corresponding to each $\sigma \in \mathcal{S}$ and scalar function $v \in C_c^\infty(\mathbb{R}^n)$ we define

$$E(\sigma, v) := \frac{1}{2} \int_{\mathbb{R}^n} |Dv|^2 \, d\sigma - \int_{\mathbb{R}^n} v \, d(\mu^+ - \mu^-). \tag{10.1}$$

Then

$$E(\sigma) := \inf\{E(\sigma, v) \mid v \in C_c^\infty(\mathbb{R}^n)\} \tag{10.2}$$

represents the negative of the Joule heating (= energy dissipation) corresponding to the given conductivity. If there exists $v \in C_c^\infty(\mathbb{R}^n)$ giving the minimum, then

$$-\operatorname{div}(\sigma Dv) = \mu^+ - \mu^- \quad \text{in } \mathbb{R}^n; \tag{10.3}$$

that is,

$$\int_{\mathbb{R}^n} Dv \cdot Dw \, d\sigma = \int_{\mathbb{R}^n} w \, d(\mu^+ - \mu^-)$$

for each $w \in C_c^\infty(\mathbb{R}^n)$. We can interpret

$$\begin{cases} v = \text{ electrostatic potential, } \mathbf{e} = -Dv = \text{ electric field,} \\ \mathbf{j} = \sigma\mathbf{e} = \text{ current density (by Ohm's law).} \end{cases}$$

We now ask: Can we find $\sigma^* \in \mathcal{S}$ to *maximize $E(\sigma)$*? In other words, is there an optimal way to arrange the given amount of conductivity material so as to *minimize* the heating?

Following Bouchitte–Buttazzo–Seppechere [B-B-S], we introduce the related Monge–Kantorovich dual problem of maximizing

$$\int_{\mathbb{R}^n} u \, d(\mu^+ - \mu^-) \tag{10.4}$$

among all $u : \mathbb{R}^n \to \mathbb{R}$, with

$$|u(x) - u(y)| \le 1 \quad (x, y \in \mathbb{R}^n). \tag{10.5}$$

Lemma 10.1 (i) *There exists u^* solving (10.4), (10.5).*

(ii) *Furthermore, there exists $\alpha^* \in \mathcal{A}$ such that*

$$\begin{cases} -\operatorname{div}(\alpha^* D u^*) = \mu^+ - \mu^- \\ |Du^*| = 1 \quad \alpha^* \ a.e. \end{cases} \tag{10.6}$$

This is clearly a generalization of our work in §9. Note carefully that since α^* is merely a Radon measure, and thus may be in part singular with respect to n-dimensional Lebesgue measure, care is needed in interpreting the PDE in (10.6): see [B-B-S].

Now set

$$\begin{cases} \sigma^* := (\alpha^*(\mathbb{R}^n))^{-1}\alpha^* \\ v^* := \alpha^*(\mathbb{R}^n)u^* \end{cases} \tag{10.7}$$

Then $\sigma^* \in \mathcal{S}$ and

$$-\operatorname{div}(\sigma^* D v^*) = \mu^+ - \mu^- \quad \text{in } \mathbb{R}^n. \tag{10.8}$$

We claim now that

$$E(\sigma^*) = \max_{\sigma \in \mathcal{S}} E(\sigma). \tag{10.9}$$

For this we invoke another duality principle, namely

$$E(\sigma) = \sup\left\{ -\frac{1}{2} \int_{\mathbb{R}^n} |\mathbf{e}|^2 \, d\sigma \mid \operatorname{div}(\sigma\mathbf{e}) = \mu^+ - \mu^- \right\}. \tag{10.10}$$

Now if u satisfies (10.5) and $\operatorname{div}(\sigma\mathbf{e}) = \mu^+ - \mu^-$, then

$$\begin{aligned} \int_{\mathbb{R}^n} u \, d(\mu^+ - \mu^-) &= -\int_{\mathbb{R}^n} Du \cdot \mathbf{e} \, d\sigma \\ &\leq \left(\int_{\mathbb{R}^n} |Du|^2 \, d\sigma \right)^{1/2} \left(\int_{\mathbb{R}^n} |\mathbf{e}|^2 \, d\sigma \right)^{1/2} \\ &\leq \left(\int_{\mathbb{R}^n} |\mathbf{e}|^2 \, d\sigma \right)^{1/2}, \end{aligned}$$

since $|Du| \leq 1$ and $\sigma(\mathbb{R}^n) = 1$. Taking the suprema over all u, we deduce

$$-\frac{1}{2} \int_{\mathbb{R}^n} |\mathbf{e}|^2 \, d\sigma \leq -\frac{1}{2}\alpha^*(\mathbb{R}^n)^2.$$

In light of (10.10), then

$$E(\sigma) \leq -\frac{1}{2}\alpha^*(\mathbb{R}^n). \tag{10.11}$$

But since $|Du^*| = 1$ σ^* a.e., we compute for $\mathbf{e}^* := -Dv^*$ that

$$E(\sigma^*) \geq -\frac{1}{2} \int_{\mathbb{R}^n} |\mathbf{e}^*|^2 \, d\sigma^* = -\frac{1}{2}\alpha^*(\mathbb{R}^n)^2.$$

According then to (10.11), σ^* is optimal.

Remark. Results strongly related to these are to be found in earlier work of Iri [I] and Strang [S1], [S2]. □

11 Application: Sandpile models

As a completely different class of applications, we next introduce some physical "sandpile" models evolving in time, for which we can identify a Monge–Kantorovich mass transfer mechanism on a "fast" time scale. In the following examples we regard u as the height function of our sandpiles: the constraint

$$|Du| \leq 1 \tag{11.1}$$

is everywhere imposed, and has the physical meaning that the sand cannot remain in equilibrium if the slope anywhere exceeds the angle of repose $\pi/4$. We will later reinterpret (11.1) as the Monge–Kantorovich constraint: the interplay of these interpretations animates much of the following. The exposition is based in part upon the interesting papers of Prigozhin [P1], [P2], [P3], and also upon [E-F-G], [A-E-W], [E-R].

11.1 Growing sandpiles

As a simple preliminary model, suppose that the function $f \geq 0$ is a source term, representing the rate that sand is added to a sandpile, whose initial height is zero. We then have $u_t = f$ in any region where the constraint (11.1) is active. But if adding more sand at some location would break the constraint, we may imagine the newly added sand particles "to roll downhill instantly", coming to rest at new sites where their addition maintains the constraint.

We propose as a model for the resulting evolution:

$$\begin{cases} f - u_t \in \partial I_\infty[u] & (t > 0) \\ u = 0 & (t = 0), \end{cases} \tag{11.2}$$

the functional $I_\infty[\cdot]$ defined in §9. The interpretation is that at each moment of time, the mass $d\mu^+ = f^+(\cdot, t)dx$ is instantly and optimally transported downhill by the potential $u(\cdot, t)$ into the mass $d\mu^- =$

$u_t(\cdot, t)dy$. In other words *the height function $u(\cdot, t)$ of the sandpile is deemed also to be the potential generating the Monge–Kantorovich reallocation of $f^+ dx$ to $u_t dy$.* This requirement forces the dynamics (11.2).

Example: Interacting sandcones. Consider, for example, the case that mass is added only at fixed sites:

$$f = \sum_{k=1}^{m} f_k(t)\delta_{d_k}, \qquad (11.3)$$

where $f_k > 0$. In this case we expect the height function u to be the union of interacting *sandcones*:

$$u(x, t) = \max\{0, z_1(t) - |x - d_1|, \dots, z_m(t) - |x - d_m|\}. \qquad (11.4)$$

Owing to conservation of mass, we expect the cone heights $\mathbf{z}(t) = (z_1(t), \dots, z_m(t))$ to solve the coupled system of ODE

$$\begin{cases} \dot{z}_k(t) = \dfrac{f_k(t)}{|D_k(t)|} & (t \geq 0) \\ z_k(0) = 0 \end{cases} \qquad (11.5)$$

for $k = 1, \dots, m$, where $|D_k(t)|$ denotes the measure of the set $D_k(t) \subset \mathbb{R}^n$ on which the k-th cone determines u:

$$D_k(t) = \{x \in \mathbb{R}^n \mid z_k(t) - |x - d_k| > 0, \ z_l(t) - |x - d_l| \text{ for } l \neq k\}. \qquad (11.6)$$

These ODE were originally derived by Aronsson [AR1].

Let us confirm that (11.4)–(11.6) give the solution of (11.2), in the case of point sources (11.3). To check (11.2) we must show at each time $t \geq 0$ that

$$I_\infty[u] + (f - u_t, v - u)_{L^2} \leq I_\infty[v] \qquad (11.7)$$

for all $v \in L^2(\mathbb{R}^n)$. Now $I_\infty[u(\cdot, t)] = 0$, and if $I_\infty[v] = +\infty$, then (11.7) is obvious. So we may as well assume $I_\infty[v] = 0$; that is,

$$|Dv| \leq 1 \text{ a.e.} \qquad (11.8)$$

The problem now is to show that

$$(f - u_t(\cdot, t), v - u(\cdot, t))_{L^2} \leq 0. \qquad (11.9)$$

Owing to (11.3) and (11.4), the term on the left means

$$\sum_{k=1}^{m} f_k(t)(v(d_k) - u(d_k, t)) - \sum_{k=1}^{\infty} \int_{D_k(t)} \dot{z}_k(t)(v(x) - u(x, t)) \, dx.$$

Consequently the ODE (11.5) means that we must show

$$\sum_{k=1}^{m} f_k(t) \fint_{D_k(t)} v(d_k) - v(x) \, dx \leq \sum_{k=1}^{m} f_k(t) \fint_{D_k(t)} u(d_k, t) - u(x, t) \, dx, \tag{11.10}$$

the slash through the integral signs denoting average. But (11.10) is easy: in light of (11.8) and (11.4), we have

$$v(d_k) - v(x) \leq |d_k - x| = u(d_k, t) - u(x, t)$$

on $D_k(t)$.   □

11.2   Collapsing sandpiles

The dynamics (11.2) model "surface flows" for sandpiles: once a sand grain is added, rolls downhill and comes to rest, it never again moves. It is therefore interesting to modify this model to allow for "avalanches".

In view of the approximation in §9 of the term $\partial I_\infty[u]$ by the p-Laplacian operator $-\mathrm{div}(|Du|^{p-2}Du)$, we propose now to investigate the limiting behavior as $p \to \infty$ of the quasilinear parabolic problem

$$\begin{cases} u_{p,t} - \mathrm{div}(|Du_p|^{p-2}Du_p) = 0 & \text{in } \mathbb{R}^n \times (0, \infty) \\ u_p = g & \text{on } \mathbb{R}^n \times \{t = 0\}. \end{cases} \tag{11.11}$$

Here $g : \mathbb{R}^n \to \mathbb{R}$ is the given initial height of a sandpile, satisfying the *instability condition*

$$L := \sup_{\mathbb{R}^n} |Dg| > 1. \tag{11.12}$$

For large p, the PDE in (11.11) supports very fast diffusion in regions where $|Du_p| > 1$ and very slow diffusion in regions where $|Du_p| < 1$. We expect therefore that for $p >> 1$ the solutions u_p of (11.11) will rapidly rearrange its mass to achieve the stability condition. As the initial profile is unstable, we expect there to be a short time period during which there is rapid mass flow, followed by a time in which u_p is practically unchanging.

Indeed, simple estimates suffice to show that there exists a function $u = u(x)$ with

$$|Du| \leq 1 \quad \text{in } \mathbb{R}^n \tag{11.13}$$

such that

$$u_{p_k} \to u \quad \text{uniformly on compact subsets of } \mathbb{R}^n \times (0, \infty) \tag{11.14}$$

for some sequence $p_k \to \infty$. We call u the *collapsed profile* and our problem is to understand how the initial, unstable profile g rearranges itself into u.

To understand the mapping $g \mapsto u$, we rescale the PDE (11.11) to stretch out the initial time layer of rapid mass motion. For this, set

$$v_p(x, t) := t u_p\left(x, \frac{t^{p-1}}{p-1}\right) \quad (x \in \mathbb{R}^n,\ t > 0) \tag{11.15}$$

and write $\tau := L^{-1}$. It turns out then that

$$v_{p_k} \to v \quad \text{uniformly on } \mathbb{R}^n \times [\tau, 1],$$

where

$$\begin{cases} \dfrac{v}{t} - v_t \in \partial I_\infty[v] & (\tau \leq t \leq 1) \\ v = h & (t = \tau) \end{cases} \tag{11.16}$$

for $h := \frac{1}{L} g$. Furthermore the collapsed profile is

$$u = v(\cdot, 1). \tag{11.17}$$

In summary, *our procedure for calculating the collapse $g \mapsto u$ is to define h as above, solve the evolution (11.16) and then set $u := v(\cdot, 1)$.*

Now (11.16) is interpreted, as above, as an evolution in which the mass $d\mu^+ = \frac{v}{t} dx$ is instantly and optimally rearranged by the potential v into $d\mu^- = v_t dy$. Again we have a Monge–Kantorovich mass reallocation occurring on a fast time scale, which thereby generates the dynamics (11.16).

Example: Collapse of a convex cone. As an application let us take Γ_τ to be the boundary of an open convex region $U_\tau \subset \mathbb{R}^2$. Assume

$$g(x) := \begin{cases} L \text{ dist}(x, \Gamma_\tau) & x \in U_\tau \\ 0 & \text{otherwise} \end{cases}$$

is the height of the initial, unstable sandpile. How does it collapse into equilibrium? Following the procedure above, we look to the evolution

$$
\begin{cases}
\dfrac{v}{t} - v_t \in \partial I_\infty[v] & (\tau \leq t \leq 1) \\
v = h & (t = \tau),
\end{cases}
\tag{11.18}
$$

for

$$
h(x) := \begin{cases}
\mathrm{dist}(x, \Gamma_\tau) & x \in U_\tau \\
0 & \text{otherwise.}
\end{cases}
$$

We next guess that our solution v has the form

$$
v(x, t) = \begin{cases}
\mathrm{dist}(x, \Gamma_t) & x \in U_t \\
0 & \text{otherwise,}
\end{cases}
$$

where Γ_t is the boundary of an open set U_t. In other words, we conjecture that v is the distance function to a moving family of surfaces $\{\Gamma_t\}_{\tau \leq t \leq 1}$.

We next employ techniques from Monge–Kantorovich theory to derive a geometric law of motion for the surfaces $\{\Gamma_t\}_{\tau \leq t \leq 1}$, which describe the moving edge of the collapsing sandpile. We hereafter write for each point $y \in \Gamma_t$,

$$
\begin{cases}
\gamma = \gamma(y, t) = \text{radius of the largest disk within} \\
U_t \text{ which touches } \Gamma_t \text{ at } y.
\end{cases}
$$

Take a point $x \in U_t$ with a unique closest point $y \in \Gamma_t$. Let $\kappa :=$ curvature of Γ_t at y and $R := \frac{1}{\kappa}$.

We may assume the segment $[x, y]$ is vertical and $y = (0, R)$. Set

$$
A_\varepsilon := \{z \mid \theta(z, e_2) < \varepsilon, \ R - \gamma \leq |z| \leq R\},
$$

where $\theta(z, e_2)$ denotes the angle between z and $e_2 = (0, 1)$. Now we interpret (11.18) as saying that the mass $d\mu^+ = \frac{v}{t} dx$ is transferred in the direction $-Dv$, to $d\mu^- = v_t dy$. This transfer, restricted to the set A_ε, forces the (approximate) *detailed mass balance* relation

$$
\int_{A_\varepsilon} \frac{v}{t} \, dx \approx \int_{A_\varepsilon} v_t \, dy,
\tag{11.19}
$$

according to Lemma 9.3. Now divide both sides by $|A_\varepsilon|$ and send $\varepsilon \to 0$. A calculation shows

$$
\lim_{\varepsilon \to 0} \fint_{A_\varepsilon} \frac{v}{t} \, dx = \frac{\gamma}{3t} \left(\frac{3 - 2\kappa\gamma}{2 - \kappa\gamma} \right)
$$

and, since $v_t = V(=$ outward normal velocity of Γ_t at y) along the ray through y,

$$\lim_{\varepsilon \to 0} \fint_{A_\varepsilon} v_t \, dy = V.$$

We derive therefore the *nonlocal geometric law of motion*

$$V = \frac{\gamma}{3t} \left(\frac{3 - 2\kappa\gamma}{2 - \kappa\gamma} \right) \quad \text{on } \Gamma_t \tag{11.20}$$

for the moving surfaces $\{\Gamma_t\}_{\tau \le t \le 1}$. (See Feldman [F] for a rigorous analysis of (11.20).)

11.3   A stochastic model

A variant of our growing sandpile evolution, discussed in §11.1, arises a rescaled continuum limit of the following stochastic model. Consider the lattice $\mathbb{Z}^2 \subseteq \mathbb{R}^2$ as subdividing the plane into unit squares. We introduce a discrete model for a "sandpile", as a stack of unit cubes resting on the plane, each column of cubes above a unit square. At each moment the configuration must be *stable*, which means that the heights of any two adjacent columns of cubes can differ by at most one. (A given column has four adjacent columns, in the coordinate directions.)

We image additional cubes being added randomly to the top of columns. If the new pile is stable, the new cube remains in place. Otherwise it "falls downhill", until it reaches a stable position.

What happens in the scaled continuum limit, when we take more and more, smaller and smaller cubes added faster and faster?

We make precise our model, generalizing to $\mathbb{R}^n$. Let us write $\mathbf{i} = (i_1, \ldots, i_n)$ for a typical site in $\mathbb{Z}^n$ and say two sites $\mathbf{i}, \mathbf{j}$ are *adjacent*, written $\mathbf{i} \sim \mathbf{j}$, if

$$\max_{1 \le k \le n} |i_k - j_k| = 1.$$

A (*stable*) *configuration* is a mapping $\eta : \mathbb{Z}^n \to \mathbb{Z}$ such that

$$\begin{cases} |\eta(\mathbf{i}) - \eta(\mathbf{j})| \le 1 \text{ if } \mathbf{i} \sim \mathbf{j} \\ \text{and } \eta \text{ has bounded support.} \end{cases}$$

The *state space* S is the collection of all configurations.

We introduce as follows a Markov process on S. Given $\eta \in S$ and $\mathbf{i} \in \mathbb{Z}^n$, write

$$\begin{aligned} \Gamma(\mathbf{i}, \eta) := \quad &\{\mathbf{j} \in \mathbb{Z}^n \mid \text{there exist sites } \mathbf{i} = \mathbf{i}_1 \sim \mathbf{i}_2 \sim \cdots \sim \mathbf{i}_m = \mathbf{j} \\ &\text{with } \eta(\mathbf{i}_{l+1}) = \eta(\mathbf{i}_l) - 1 \ (l = 1, \ldots, m-1), \\ &\eta(\mathbf{k}) \ne \eta(\mathbf{j}) - 1 \text{ for all } \mathbf{k} \sim \mathbf{j}\}. \end{aligned} \tag{11.21}$$

Thus $\Gamma(\mathbf{i}, \eta)$ is the set of sites $\mathbf{j}$ at which a cube newly added to $\mathbf{i}$ can come to rest. Assign to each $\mathbf{j} \in \Gamma(\mathbf{i}, \eta)$ a number

$$0 \le p(\mathbf{i}, \mathbf{j}, \eta) \le 1$$

such that

$$\sum_{\mathbf{j} \in \Gamma(\mathbf{i}, \eta)} p(\mathbf{i}, \mathbf{j}, \eta) = 1, \tag{11.22}$$

and think of $p(\mathbf{i}, \mathbf{j}, \eta)$ as the probability that a cube added to $\mathbf{i}$ will fall downhill and come to rest at $\mathbf{j}$. Finally set

$$c(\mathbf{j}, \eta, t) := \sum_{\mathbf{i} : \mathbf{j} \in \Gamma(\mathbf{i}, \eta)} p(\mathbf{i}, \mathbf{j}, \eta) f\left(\frac{\mathbf{i}}{\mathbf{N}}, \frac{t}{N}\right) \tag{11.23}$$

where $f : \mathbb{R}^n \times [0, \infty) \to [0, \infty)$ is a given function, the *source density*. Then $c(\mathbf{j}, \eta, t)$ is the rate at which cubes are coming to rest at $\mathbf{j}$.

We introduce the *infinitesimal generator* $\mathcal{L}_t$ of our Markov process by taking any $F : S \to \mathbb{R}$ and defining then

$$(\mathcal{L}_t F)(\eta) := \sum_{\mathbf{j} \in \mathbb{Z}^n} c(\mathbf{j}, \eta, t)(F(\eta^{\mathbf{j}}) - F(\eta)), \tag{11.24}$$

where

$$\eta^{\mathbf{j}}(\mathbf{i}) := \begin{cases} \eta(\mathbf{i}) + 1 & \text{if } \mathbf{i} = \mathbf{j} \\ \eta(\mathbf{i}) & \text{if } \mathbf{i} \ne \mathbf{j}. \end{cases}$$

The formula (11.24) encodes the foregoing probabilistic interpretation.

Let $\{\eta(\cdot, t)\}_{t \geq 0}$ denote the inhomogeneous Markov process on S generated by $\{\mathcal{L}_t\}_{t \geq 0}$, with $\eta(\cdot, 0) \equiv 0$. Thus $\eta(\mathbf{i}, t)$ is the (random) height of the pile of cubes at site $\mathbf{i}$, time $t \geq 0$.

We intend now to rescale, taking now cubes of side $\frac{1}{N}$ added on a time scale multiplied by N. We construct therefore the rescaled process $\frac{1}{N}\eta([xN], tN)$, where $x \in \mathbb{R}^n$, $t \geq 0$, and inquire what happens as $N \to \infty$.

Theorem 11.1 *For each $t \geq 0$, we have*

$$E\left(\sup_{x \in \mathbb{R}^n} \left|\frac{1}{N}\eta([xN], tN) - u(x, t)\right|\right) \to 0 \qquad (11.25)$$

as $N \to \infty$, where u is the unique solution of the evolution

$$\begin{cases} f - u_t \in \partial \hat{I}[u] & (t = 0) \\ u = 0 & (t = 0). \end{cases} \qquad (11.26)$$

Here

$$\hat{I}[v] := \begin{cases} 0 & \text{if } v \in \hat{\mathbb{K}} \\ +\infty & \text{otherwise} \end{cases}$$

for

$$\hat{\mathbb{K}} := \{v \in L^2(\mathbb{R}^n) \mid v \text{ is Lipschitz}, |v_{x_i}| \leq 1 \text{ a.e. } (i = 1, \ldots, n)\}.$$

Consequently, the continuum dynamics (11.26), while similar to (11.2), differ by "remembering" the anisotropic structure of the lattice.

The proof in [E-R] is too complicated to reproduce here: the key new idea is a combinatorial lemma asserting that if η, ξ are any two configurations in S, then

$$\sum_{\mathbf{j}} c(\mathbf{j}, \eta, t)(\eta(\mathbf{j}) - \xi(\mathbf{j})) \leq \sum_{\mathbf{i}} f\left(\frac{\mathbf{i}}{\mathbf{N}}, \frac{t}{N}\right)(\eta(\mathbf{i}) - \xi(\mathbf{i})). \qquad (11.27)$$

Since the term c, describing the rate at which cubes come to rest at a given site, is a sort of rescaled, discrete analogue of u_t, (11.27) is a "microscopic" analogue of the "macroscopic" inequality

$$\int_{\mathbb{R}^n} (f - u_t)(v - u)\,dx \leq 0 \quad (v \in \hat{\mathbb{K}}).$$

But this is just, as we have seen before, another way of writing the evolution (11.26).

A further interpretation of (11.26) is that the mass $d\mu^+ = f dx$ is instantly rearranged into $d\mu^- = u_t dy$, so as to minimize the cost

$$\int_{\mathbb{R}^n} c(x, \mathbf{s}(x)) f^+(x)\, dx$$

for the l^1-distance

$$c(x, y) = |x_1 - y_1| + \cdots + |x_n - y_n|.$$

In the case of point sources, $f = \sum f_k(t)\delta_{d_k}$, the dynamics (11.26) correspond to growing, interacting *pyramids* of sand.

12   Application: Compression molding

The utterly different physical situation of *compression molding* gives rise to similar mathematics.

In this setting we consider an incompressible plastic material being squeezed between two horizontal plates.

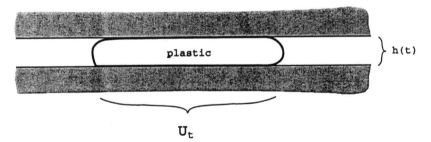

Assume the lower plate is fixed, and the plate separation at time t is $h(t)$, with $\dot{h}(t) \leq 0$, $h(T) = 0$. Let U_t denote the (approximate) projection onto $\mathbb{R}^2$ of the region occupied by the plastic at time t and write $\Gamma_t := \partial U_t$. As $t \to T$, the region U_t expands to fill the entire plane.

Following [AR2], [A-E], we introduce a highly simplified model of the physics, with the goal of tracking the air-plastic interface Γ_t for times $0 \leq t < T$. After a number of simplifying assumptions for this highly viscous flow, the relevant PDE becomes

$$\operatorname{div}(|Dp|^{\frac{1}{\sigma}-1} Dp) = \frac{\dot{h}}{h} \quad \text{in } U_t \tag{12.1}$$

where $0 \leq t < T$, p = pressure, and σ is a small constant arising in a power-law constitutive relation for the highly non-Newtonian flow.

We can rescale in time to convert to the case $T = \infty$, $\dot{h}/h \equiv -1$, in which case (12.1) reads

$$- \operatorname{div}(|Dp|^{\frac{1}{\sigma}-1}Dp) = 1 \qquad \text{in } U_t \tag{12.2}$$

for $t \geq 0$. In addition, we have the boundary conditions

$$\begin{cases} p = 0 & \text{on } \Gamma_t \\ V = |Dp|^{1/\sigma} & \text{on } \Gamma_t, \end{cases} \tag{12.3}$$

where V = outward normal velocity of Γ_t.

Now experimental results in the engineering literature suggest that for real materials the evolution of Γ_t does not much depend upon the exact choice of σ, so long as σ is small: see, for instance, [F-F-T]. We therefore propose to study the asymptotic limit $\sigma \to 0$. For this, we first change notation, and rewrite (12.2), (12.3) as

$$\begin{cases} -\operatorname{div}(|Du_p|^{p-2}Du_p) = 1 & \text{in } U_t \\ u_p = 0, \ V = |Du_p|^{p-2} & \text{on } \Gamma_t. \end{cases} \tag{12.4}$$

When $p \to \infty$, we expect in light of the calculations in §9 that $u_p \to u$, where

$$\begin{cases} -\operatorname{div}(aDu) = 1 & \text{in } U_t \\ u = 0, \ V = a & \text{on } \Gamma_t \end{cases} \tag{12.5}$$

for some nonnegative function a. The physical interpretations are now

$$\begin{cases} u = \text{pressure}, \ \mathbf{v} = -aDu = \text{velocity}, \\ a = \text{speed}. \end{cases}$$

We can further reinterpret (12.5) as the evolution

$$\begin{cases} w - w_t \in \partial I_\infty[u] \\ u \in \beta(w), \end{cases} \quad (t > 0) \tag{12.6}$$

where β is the multivalued mapping

$$\beta(x) = \begin{cases} [0, \infty) & \text{if } x \geq 1 \\ 0 & \text{if } 0 \leq x \leq 1 \\ (-\infty, 0] & \text{if } x \leq 0. \end{cases}$$

We interpret $w = \chi_{U_t}$, and so the initial condition for (12.6) is

$$w = \chi_{U_0} \quad (t = 0). \tag{12.7}$$

The Monge–Kantorovich interpretation is that at each moment the measure $d\mu^+ = \chi_{U_t} dx$ is being instantly and optimally rearranged to $d\mu^- = V$ times (n-1)-dimensional Hausdorff measure $\mathcal{H}^{n-1}$ restricted to Γ_t. These dynamics in turn determine the velocity V of Γ_t.

It is not hard to check that

$$u(x,t) = \begin{cases} \text{dist}(x, \Gamma_t) & x \in U_t \\ 0 & \text{otherwise;} \end{cases}$$

so that the pressure in this asymptotic model is just the distance to the boundary.

Using detailed mass balance arguments somewhat like those in §11 we can further show that

$$V = \gamma \left(1 - \frac{\kappa \gamma}{2}\right) \quad \text{on } \Gamma_t \tag{12.8}$$

where $\gamma =$ radius of largest disk within U_t touching Γ_t at y and $\kappa =$ curvature. Feldman [F] has rigorously analyzed this geometric flow that characterizes the spread of the plastic.

Appendix

13 Finite-dimensional linear programming

To motivate some functional analytic considerations in §1, we record here some facts about linear programming. A nice reference is Ekeland–Turnbull [E-T].

Notation. If $x = (x_1, \ldots, x_N) \in \mathbb{R}^N$, we write

$$x \geq 0$$

to mean $x_k \geq 0$ $(k = 1, \ldots, N)$. □

Assume we are given $c \in \mathbb{R}^N$, $b \in \mathbb{R}^M$, and an $M \times N$ matrix A.

The *primal linear programming problem* is to find $x^* \in \mathbb{R}^N$ so as to

(P)
$$\begin{cases} \text{minimize } c \cdot x, \text{ subject to the} \\ \text{constraints } Ax \geq b, \ x \geq 0. \end{cases}$$

The *dual problem* is then to find $y^* \in \mathbb{R}^M$ so as to

(D) $\qquad \begin{cases} \text{maximize } b \cdot y, \text{ subject to the} \\ \text{constraints } A^T y \leq c, \ y \geq 0. \end{cases}$

Assume $x^* \in \mathbb{R}^N$ solves (P), $y^* \in \mathbb{R}^M$ solves (D). We may then regard y^* as the Lagrange multiplier for the constraints in (P), and, likewise, x^* as the Lagrange multiplier for (D). Furthermore we have the saddle point relation:

$$c \cdot x^* = b \cdot y^*; \qquad (13.1)$$

that is,

$$\min\{c \cdot x \mid Ax \geq b, \ x \geq 0\} = \max\{b \cdot y \mid A^T y \leq c, \ y \geq 0\}. \qquad (13.2)$$

An example. Suppose we are given nonnegative numbers c_{ij}, μ_i^+, μ_j^- $(i = 1, \ldots, n; \ j = 1, \ldots, m)$, with

$$\sum_{i=1}^{n} \mu_i^+ = \sum_{j=1}^{m} \mu_j^-,$$

and we are asked to find μ_{ij}^* $(i = 1, \ldots, n; \ j = 1, \ldots, m)$, so as to

$$\text{minimize } \sum_{i=1}^{n} \sum_{j=1}^{m} c_{ij} \mu_{ij}, \qquad (13.3)$$

subject to the constraints

$$\sum_{j=1}^{m} \mu_{ij} = \mu_i^+, \ \sum_{i=1}^{n} \mu_{ij} = \mu_j^-, \ \mu_{ij} \geq 0 \ (i = 1, \ldots, n; \ j = 1, \ldots, m). \qquad (13.4)$$

This is a problem of the form (P), with

$$\begin{cases} N = nm, \ M = n + m, \ x = (\mu_{11}, \mu_{12}, \ldots, \mu_{1m}, \mu_{21}, \ldots, \mu_{nm}), \\ c = (c_{11}, \ldots, c_{1m}, c_{21}, \ldots, c_{nm}), \ b = (\mu_1^+, \ldots, \mu_n^+, \mu_1^-, \ldots, \mu_m^-), \end{cases}$$

and A the $(n + m) \times nm$ matrix

$$\begin{matrix} n \text{ rows} \left\{ \vphantom{\begin{matrix} 1 \\ 0 \\ 0 \end{matrix}} \right. \\ \\ m \text{ rows} \left\{ \vphantom{\begin{matrix} e_1 \\ e_2 \\ e_n \end{matrix}} \right. \end{matrix} \begin{pmatrix} \mathbb{1} & 0 & \cdots & 0 \\ 0 & \mathbb{1} & \cdots & 0 \\ 0 & 0 & \cdots & \mathbb{1} \\ \cdots & \cdots & \cdots & \cdots \\ e_1 & e_1 & \cdots & e_1 \\ e_2 & e_2 & \cdots & e_2 \\ e_n & e_n & \cdots & e_n \end{pmatrix}$$

Above the middle horizontal dotted line, each entry is a row vector in $\mathbb{R}^m$ and $\mathbb{1} = (1,\ldots,1)$, $0 = (0,0,\ldots,0)$. Below the dotted line each entry is a row vector in $\mathbb{R}^n$ and $e_i = (0,0,\ldots,1,\ldots,0)$, the 1 in the i-th slot $(i = 1,\ldots,n)$.

Now write $y = (u_1,\ldots,u_n,v_1,\ldots,v_m) \in \mathbb{R}^{n+m}$. Employing the explicit form of the matrix A, we translate the dual problem (D) for the case at hand to read

$$\text{maximize} \quad \sum_{i=1}^{n} u_i \mu_i^+ + \sum_{j=1}^{m} v_j \mu_j^-, \tag{13.5}$$

subject to the constraints

$$u_i + v_j \le c_{ij}, \ u_i \ge 0, \ v_j \ge 0 \ (i = 1,\ldots,n; \ j = 1,\ldots,m). \tag{13.6}$$

The Monge–Kantorovich mass transfer problem is a continuum version of (13.3), (13.4) and its dual problem a continuum version of (13.5), (13.6): see §1. □

References

[A]   P. Appell, Le probleme geometrique des déblais et remblais, Memor. des Sciences Math., Acad. de Sciences de Paris, Gauthier Villars **27** (1928), 1–34.

[AR1]   G. Aronsson, A mathematical model in sand mechanics, SIAM J. Applied Math, **22** (1972), 437–458.

[AR2]   G. Aronsson, Asymptotic solutions of a compression molding problem, preprint, Department of Mathematics, Linkoping University.

[A-E]   G. Aronsson and L. C. Evans, An asymptotic model for compression molding, to appear.

[A-E-W] G. Aronsson, L. C. Evans and Y. Wu, Fast/slow diffusion and growing sandpiles, J. Differential Equations **131** (1996), 304–335.

[Ba]   F. Barthe, Inégalités fonctionnelles et géométriques obtenues par transport des mesures, Thesis, Université de Marne–la–Vallée, 1997.

[Be1]   J–D. Benamou, Transformations conservant la mesure, mécanique des fluides incompressibles et modèle semi-géostrophique en météorologie, Thesis, Université de Paris IX, 1992.

[Be2]   J–D. Benamou, A domain decomposition method for the polar factorization of vector–valued mappings, SIAM J Numerical Analysis **32** (1995), 1808–1838.

[Be-B1] J–D. Benamou and Y. Brenier, Weak existence for the semi–geostrophic equations formulated as a coupled Monge–Ampere/transport problem, preprint.

[Be-B2] J–D. Benamou and Y. Brenier, The optimal time–continuous mass transport problem and its augmented Lagrangian numerical resolution, preprint.

[B-B-M] T. Bhattacharya, E. DiBenedetto, and J. Manfredi, Limits at $p \to \infty$ of $\Delta_p u = f$ and related extremal problems, Rend. Sem. Mat. Univ. Pol. Torino, Fascicolo Speciale (1989).

[B-B-S] G. Bouchitte, G. Buttazzo and P. Seppechere, Shape optimization solutions via Monge–Kantorovich equation, CRAS **324** (1997), 1185–1191.

[B1]   Y. Brenier, Décomposition polaire et réarrangement monotone des champs de vecteurs, CRAS **305** (1987), 805–808.

[B2]   Y. Brenier, Polar factorization and monotone rearrangement of vector-valued functions, Comm. Pure Appl. Math. **44** (1991), 375–417.

[B3]   Y. Brenier, The dual least action problem for an ideal, compressible fluid, Arch. Rat. Mech. Analysis **122** (1993), 323–351.

[B4]   Y. Brenier, A geometric presentation of the semi–geostrophic equations, preprint.

[C1]   L. Caffarelli, A localization property of viscosity solutions of the Monge–Ampere equaqtion, Annals of Math. **131** (1990), 129–134.

[C2]   L. Caffarelli, Interior $W^{2,p}$ estimates for solutions of the Monge–Ampere equation, Annals of Math. **131** (1990), 135–150.

[C3]   L. Caffarelli, Some regularity properties of solutions to the Monge–Ampere equation, Comm. in Pure Appl. Math. **44** (1991), 965–969.

[C4]   L. Caffarelli, Allocation maps with general cost functions, in *Partial Differential Equations with Applications* (ed. by Talenti), 1996, Dekker.

[C5]   L. Caffarelli, The regularity of mappings with a convex potential, J. Amer. Math. Soc. **5** (1992), 99–104.

[C6]   L. Cafarelli, Boundary regularity of maps with convex potentials, Comm. Pure Appl. Math. **45** (1992), 1141–1151.

[C7]   L. Cafarelli, Boundary regularity of maps with convex potentials II, Annals of Math. **144** (1996), 453–496.

[C-P]   A. Cellina and S. Perrotta, On the validity of the Euler–Lagrange equations, preprint.

[C-S]   S. Chynoweth and M. J. Sewell, Dual variables in semigeostrophic theory, Proc. Royal Soc. London A **424** (1989), 155–186.

[C]   M. J. P. Cullen, Solutions to a model of a front forced by deformation, Quart. J. Royal Meteor. Soc. **109** (1983), 565–573.

[C-N-P]   M. J. P. Cullen, J. Norbury, and R. J. Purser, Generalized Lagrangian solutions for atmospheric and oceanic flows, SIAM J. Appl. Math. **51** (1991), 20–31.

[C-P1]   M. J. P. Cullen and R. J. Purser, An extended Lagrangian theory of semigeostrophic frontogenesis, J. of the Atmospheric Sciences **41** (1984), 1477–1497.

[C-P2]   M. J. P. Cullen and R. J. Purser, A duality principle in semigeostgrophic theory, J. of the Atmospheric Sciences **44** (1987), 3449–3468.

[C-P3]   M. J. P. Cullen and R. J. Purser, Properties of the Lagrangian semigeostrophic equations, J. of the Atmospheric Sciences **46** (1989), 2684–2697.

[D-M] B. Dacorogna and J. Moser, On a partial differential equation involving the Jacobian determinant, Ann. Inst. H. Poincare **7** (1990), 1–26.

[D] C. Dupin, *Applications de Geometrie et de Mechanique*, Bachelier, Paris, 1822.

[E-T] I. Ekeland and T. Turnbull, *Infinite-dimensional optimization and convexity*, Chicago Lectures in Mathematics, Univ. of Chicago Press, 1983.

[E-G1] L. C. Evans and W. Gangbo, Differential equations methods in the Monge–Kantorovich mass transfer problem, Memoirs American Math. Society, to appear.

[E-G2] L. C. Evans and R. Gariepy, *Measure Theory and Fine Properties of Functions*, CRC Press, 1992. (An errata sheet for the first printing of this book is available through the math.berkeley.edu website.)

[E-F-G] L. C. Evans, M. Feldman and R. Gariepy, Fast/slow diffusion and collapsing sandpiles, J. Differential Equations **137** (1997), 166–209.

[E-R] L. C. Evans and F. Rezakhanlou, A stochastic model for sandpiles and its continuum limit, to appear.

[F] M. Feldman, Variational evolution problems and nonlocal geometric motion, forthcoming.

[F-F-T] F. Folgar, C.-C. Lee and C. L. Tucker, Simulation of compression molding for fiber-reinforced thermosetting polymers, Trans. ASME J. of Eng. for Industry **106** (1984), 114–125.

[G] W. Gangbo, An elementary proof of the polar factorization of vector-valued functions, Arch. Rat. Math. Analysis **128** (1994), 381–399.

[G-M1] W. Gangbo and R. McCann, Optimal maps in Monge's mass transport problem, C.R. Acad. Sci. Paris **321** (1995), 1653–1658.

[G-M2] W. Gangbo and R. McCann, The geometry of optimal transport, Acta Mathematica **177**, (1996), 113–161.

[I]   M. Iri, Theory of flows in continua as approximation to flows in networks, in *Survey of Math Programming* (ed. by Prekopa), North-Holland, 1979.

[J]   U. Janfalk, On certain problems concerning the p-Laplace operator, Linköping Studies in Science and Technology, Dissertation #326, Linköping University, Sweden, 1993.

[J-K-O1] R. Jordan, D. Kinderlehrer, and F. Otto, The variational formulation of the Fokker–Planck equation, preprint.

[J-K-O2] R. Jordan, D. Kinderlehrer, and F. Otto, The route to stability through Fokker–Planck dynamics, Proc. First US-China Conference on Differential Equations and Applications.

[Je]   R. Jensen, Uniqueness of Lipschitz extensions minimizing the sup-norm of the gradient, Arch. Rat. Mech. Analysis **123** (1993), 51–74.

[K1]   L. V. Kantorovich, On the transfer of masses, Dokl. Akad. Nauk. SSSR **37** (1942), 227–229 (Russian).

[K2]   L. V. Kantorovich, On a problem of Monge, Uspekhi Mat. Nauk. **3** (1948), 225–226.

[MC1]   R. McCann, A convexity theory of interacting gases, Advances in Mathematics **128** (1997), 153–179.

[MC2]   R. McCann, Existence and uniqueness of monotone measure-preserving maps, Duke Math. J. **80** (1995), 309–323.

[MC3]   R. McCann, Equilibrium shapes for planar crystals in an external field, preprint.

[MC4]   R. McCann, Exact solutions to the transportation problem on the line, preprint.

[M]   G. Monge, *Memoire sur la Theorie des Déblais et des Remblais*, Histoire de l'Acad. des Sciences de Paris, 1781.

[O1]   F. Otto, Dynamics of labyrinthine pattern formation in magnetic fields, preprint.

[O2]   F. Otto, letter.

[P1]   L. Prigozhin, A variational problem of bulk solid mechanics and free-surface segratation, Chemical Eng. Sci. **78** (1993), 3647–3656.

[P2]   L. Prigozhin, Sandpiles and river networks: extended systems with nonlocal interactions, Phys. Rev E **49** (1994), 1161–1167.

[P3]   L. Prigozhin, Variational model of sandpile growth, European J. Appl. Math. **7** (1996), 225–235.

[R]   S. T. Rachev, The Monge–Kantorovich mass transference problem and its stochastic applications, Theory of Prob. and Appl. **29** (1984), 647–676.

[Rk]   T. Rockafeller, Characterization of the subdifferentials of convex functions, Pacific Journal of Math. **17** (1966), 497–510.

[Ro]   V. A. Rohlin, On the fundamental ideas of measure theory, American Math. Society, Translations in Math. **71** (1952).

[S1]   G. Strang, Maximal flow through a domain, Math. Programming **26** (1983), 123–143.

[S2]   G. Strang, L^1 and L^∞ approximations of vector fields in the plane, in *Lecture Notes in Num. Appl. Analysis* **5** (1982), 273–288.

[St]   D. Struik, *Lectures on Classical Differential Geometry* (2nd ed.), Dover, 1988.

[Su]   V. N. Sudakov, Geometric problems in the theory of infinite-dimensional probability distributions, Proceedings of Steklov Institute **141** (1979), 1–178.

[U]   J. Urbas, On the second boundary value problem for equations of Monge–Ampere type, J. Reine Angew. Math. **487** (1997).

[W]   J. Wolfson, Minimal Lagrangian diffeomorphisms and the Monge–Ampere equation, J. Differential Geom. **46** (1997), 335–373.

Quantum Chaos, Symmetry and Zeta Functions[1]
Lecture I: Quantum Chaos

Peter Sarnak

Department of Mathematics,
Princeton University,
Princeton, NJ 08544.
sarnak@math.princeton.edu

1 Introduction

It is perhaps a little premature to give a mathematical lecture such as this on a subject in which the proven results are rather modest. However given the attention the subject has received in the Physics literature (the results being primarily computational) and given the interesting phenomena and conjectures that have emerged, it seems worthwhile. Quantum Chaos is concerned with the study of the semiclassical limit of the quantization of a classically

---

[1] Parts of these lectures were presented at the following conferences: ICMP Brisbane 1997, Joint AMS-SAMS 1997, Supersymmetry and trace formulae (Newton Institute 1997), and Journée Arithmetique (Limoges 1997).

chaotic Hamiltonian. More precisely let $H(q,p)$ be a Hamiltonian on a $2n$-dimensional symplectic manifold. We assume that the constant energy sets $H(q,p) = $ constant are compact. Let $\widehat{H}(\hbar)$ be a quantization of the classical flow defined by H, with $\hbar \to 0$ corresponding to the semi-classical limit (we give numerous explicit examples below). The corresponding Schrödinger eigenvalue equation is

$$\widehat{H}(\hbar)\psi(\hbar) + \lambda(\hbar)\psi(\hbar) = 0 \qquad (1.1)$$

It has a discrete set of eigenvalues

$$\lambda_1(\hbar) \le \lambda_2(\hbar) \le \lambda_3(\hbar) \ldots \longrightarrow \infty \qquad (1.2)$$

with a corresponding orthonormal basis of eigenfunctions

$$\psi_1(\hbar), \psi_2(\hbar), \ldots \qquad (1.3)$$

The problem mentioned above is to investigate the fine structure of the spectrum $(\lambda_j(\hbar)$'s) or eigenfunctions $(\psi_j(\hbar))$, either as $\hbar \to 0$ (semi-classical limit) or as $j \to \infty$ (large energy limit). In most of the examples below these are equivalent. We consider two extreme cases for the classical motion:

(I) The classical flow is completely integrable — that is, there are n Poisson commuting invariants of the motion $H_1 = H, H_2, \ldots, H_n$ [1].

(II) The classical flow is chaotic by which we mean that it is ergodic on the invariant surface $H = $ constant and that it has positive Liapunov exponents, etc. [1,2].

A rich family of Hamiltonians are those coming form a Riemannian metric on a manifold X. That is, H defined on $T^*(X)$ is given in the form

$$H(x,\xi) = \frac{1}{2}\sum_{i,j} g_{ij}(x)\xi^i\xi^j \qquad (1.4)$$

where $g(x)$ is the metric on X and ξ is a cotangent vector. In this case the Hamilton flow is just the geodesic flow. The standard quantization of H is $\widehat{H} = \Delta := $ div grad, the Laplace-Beltrami operator on functions on X. In these cases we suppress $\hbar$, since the $\hbar \to 0$ limit coincides with the $\lambda_j \to \infty$ limit. Among these geodesic flows, examples of (I) above are surfaces of revolution in $\mathbb{R}^3$, for which the Clairaut Integral [2] provides a second integral of the motion. A compact manifold X of negative sectional curvature provides an archetypical example of (II) above (Hopf, Anosov [1]).

We do not discuss at all the very interesting questions about the behavior of $\psi_j(\hbar)$ as $\hbar \to 0$. See Sarnak [3], Zelditch [4] and Hurt [5] for surveys of the mathematical results concerning the eigenfunctions and Heller [6] for the numerical phenomenon of enhancement of such eigenstates on the unstable periodic orbits. Our aim here is to review the developments concerning the spectrum. In order to describe the basic Conjecture, we need to digress into the topic of random matrix theory.

2 Random Matrix Models

In the early 50's Wigner [7] suggested that the resonance lines of a heavy nucleus (their determination by analytic means being intractable) might be modeled by the spectrum of a large random matrix. He introduced the ensembles: Gaussian Orthogonal Ensemble "GOE" and Gaussian Unitary Ensemble "GUE" which are probability measures on $N \times N$ real symmetric and $N \times N$ Hermitian matrices, respectively. In the first case the measure is invariant under the action of the orthogonal group while in the second it is invariant by the unitary group. He raised the question of the local (scaled) spacings between the eigenvalues (see below for a precise definition) of members of these ensembles as $N \to \infty$. The answer was provided by Gaudin [8] and Gaudin-Mehta [9], by an ingenious use of orthogonal polynomials. Later Dyson [10] introduced his closely related circular ensembles COE, CUE and CSE which he termed the 3-fold way. These are the Riemannian Symmetric spaces (with their volume forms) $U(N)/O(N)$, $U(N)$ and $U(2N)/USp(2N)$, respectively. He investigated their local spacing statistics and showed that those of the first two agree with GOE and GUE. Recent works by Altland and Zirnbauer [11] on some quantum problems (e.g. "Chaotic Andreev Quantum Dot") and by Katz and Sarnak [12,13] on zeros of zeta functions — see Lecture 2 — show that in some finer analysis these three symmetry classes do not suffice. It appears that each of the infinite families of irreducible symmetric spaces of Cartan (see [14] and [11]) should be considered. We describe five such families which play a role in these lectures.

- $U(N)$, the compact group of $N \times N$ unitary matrices in its standard realization. With Haar measure, this is Dyson's CUE.

- $USp(2N)$, the compact group of $2N \times 2N$ unitary matrices which preserve the standard symplectic form, that is, all $A \in U(2N)$ satisfying ${}^t AJA = J$, where $J = \begin{bmatrix} 0 & I_N \\ -I_N & 0 \end{bmatrix}$.

- $SO(2N)$, the compact group of $2N \times 2N$ unitary matrices preserving the standard Euclidean inner product, that is, $A \in U(2N)$ satisfying ${}^t AA = I$ and with $\det A = 1$.

- $SO(2N+1)$, same as above but odd dimensional.

All of the above compact groups become Riemannian symmetric spaces when equipped with a bi-invariant metric and they comprise the symmetric spaces of type II, see [14].

- $U(N)/O(N)$ (this is Dyson's COE), which we realize as the symmetric, unitary $N \times N$ matrices A. The map $B \to B^t B$ identifies $U(N)/O(N)$ with these matrices and turns this into a compact symmetric space for which the corresponding volume form gives the COE. Note that this symmetric space is the compact dual of $SL_N(\mathbb{R})/SO(N)$ (or more precisely $GL_N(\mathbb{R})/O(N)$) which is Wigner's GOE.

We denote by $G(N)$ any one of the above ensembles whose members A are $N \times N$ unitary matrices and whose invariant probability measure is denoted by dA. Denote the eigenvalues of an $A \in G(N)$ by $e^{i\theta_1(A)}, e^{i\theta_2(A)}, \ldots, e^{i\theta_N(A)}$ where

$$0 \leq \theta_1(A) \leq \theta_2(A) \ldots \leq \theta_N(A) < 2\pi \tag{2.1}$$

We turn to the definitions of the local scaling spacing distributions. For $k \geq 1$ the k-th (scaling) consecutive spacing measure $\mu_k(A)$ on $[0, \infty)$ is defined to be

$$\mu_k(A)([a,b]) = \frac{\#\left\{1 \leq j \leq N | \frac{N}{2\pi}(\theta_{j+k} - \theta_j) \in [a,b]\right\}}{N} \tag{2.2}$$

Notice the factor $\frac{N}{2\pi}$ which scales the spacings so that the mean of $\mu_k(A)$ is equal to k. The pair correlation $R_2(A)$ measures the distribution between all pairs of (scaled) eigenvalues of A. For $[a,b] \subset \mathbb{R}$,

$$R_2(A)[a,b] = \frac{\#\left\{j \neq k | \frac{N}{2\pi}(\theta_j - \theta_k) \in [a,b]\right\}}{N} . \tag{2.3}$$

Higher correlations may be defined similarly (see [12]).

The behavior of the measures above as $N \to \infty$ may be studied by the method of Gaudin mentioned above. Katz and Sarnak [12] show that there are measures $\mu_k(\text{II})$ and $R_2(\text{II})$, such that for any of the four families $G(N)$ of the type II symmetric spaces above, we have

$$\lim_{N \to \infty} \int_{G(N)} \mu_k(A) dA = \mu_k(\text{II}) . \tag{2.4}$$

A "Law of large numbers" which ensures that for the typical (in measure) $A \in G(N)$, $\mu_k(A)$ and $R_2(A)$ approach $\mu_k(\text{II})$ and $R_2(\text{II})$ as $N \to \infty$:

$$\lim_{N \to \infty} \int_{G(N)} D(\mu_k(A), \mu_k(\text{II})) dA = 0 \tag{2.5}$$

where
$$D(\nu_1, \nu_2) = \sup\{|\nu_1(I) - \nu_2(I)|: \ I \subset R \text{ is an interval}\} \qquad (2.6)$$
is the Kolmogoroff-Smirnov distance between ν_1 and ν_2.
For $[a, b] \subset \mathbb{R}$,

$$\lim_{N \to \infty} \int_{G(N)} |R_2(A)[a, b] - R_2(\text{II})[a, b]| dA = 0 \qquad (2.7)$$

Since one of the ensembles of type II is $U(N)$ that is, CUE, it follows that $\mu_k(\text{II})$ and $R_2(\text{II})$ coincide with the corresponding measures for CUE (or GUE), $\mu_k(\text{GUE})$ and $R_2(\text{GUE})$. These were determined by Gaudin [8] and Dyson [10], respectively. Dyson shows that

$$R_2(\text{CUE})[a, b] = \int_a^b \left(1 - \left(\frac{\sin \pi x}{\pi x}\right)^2\right) dx \qquad (2.8)$$

while Gaudin finds an expression for the densities of $\mu_k(\text{CUE})$ in terms of Fredholm determinants of operators on $L^2[-1, 1]$, see [8,9]. The latter allows for a numerical computation of these densities and in particular their graphs are given in [9].

For $G(N) = U(N)/O(N)$, that is, COE, the corresponding limits $\mu_k(\text{COE})$ $(= \mu_k(\text{GOE}))$ were determined similarly, see Mehta [9]. They are quite different to the CUE measures. For example

$$R_2(\text{GOE})[a, b] = \int_a^b r_{\text{GOE}}(x) dx \qquad (2.9)$$

where

$$r_{\text{GOE}}(x) = 1 - \left(\frac{\sin \pi x}{\pi x}\right)^2 - \left(\frac{\sin \pi x}{\pi x}\right)' \int_x^\infty \left(\frac{\sin \pi t}{\pi t}\right) dt \qquad (2.10)$$

The results above for type II $G(N)$'s show that the spacings between **all** the eigenvalues of a typical A in such a $G(N)$ is universal as $N \to \infty$. This is in sharp contrast to the distribution of the eigenvalue closest to 1. For $k \geq 1$ let $\nu_k(G(N))$ be the distribution on $[0, \infty)$ of the k-th eigenvalue of A, as A varies over $G(N)$. That is,

$$\nu_k(G(N))[a, b] = \left|\left\{A \in G(N) \ \Big| \ \frac{\theta_k(A)N}{2\pi} \in [a, b]\right\}\right| \qquad (2.11)$$

Similarly, one defines the 1-level scaling densities (or, more generally, n-level densities) of eigenvalues near 1. For $A \in G(N)$ and $[a, b] \subset \mathbb{R}$, let

$$D_1(A)[a, b] = \#\left\{\theta(A) \ \Big| \ e^{i\theta(A)} \text{ is an eigenvalue of } A \text{ and } \frac{\theta(A)N}{2\pi} \in [a, b]\right\} \qquad (2.12)$$

The averages of D are denoted W;

$$W_1(G(N)) = \int_{G(N)} D_1(A)dA \qquad (2.13)$$

In [12] it is shown that there are measures $\nu_k(G)$ on $[0, \infty)$ (which depend on the ensemble G — see [12] for graphs of their densities) such that

$$\lim_{N \to \infty} \nu_k(G(N)) = \nu_k(G) \qquad (2.14)$$

and

$$\lim_{N \to \infty} W_1(G(N))[a, b] = \int_a^b w_1(G)(x)dx \qquad (2.15)$$

where

$$w_1(G)(x) = \begin{cases} 1 & \text{if } G = U \text{ (or SU)} \\ 1 - \frac{\sin 2\pi x}{2\pi x} & \text{if } G = \text{Sp} \\ 1 + \frac{\sin 2\pi x}{2\pi x} & \text{if } G = \text{SO (even)} \\ \delta_0 + 1 - \frac{\sin 2\pi x}{2\pi x} & \text{if } G = \text{SO (odd)} \end{cases} \qquad (2.16)$$

The above models $G(N)$ are all (essentially) irreducible symmetric spaces. The antithesis is the case of a completely reducible space such as the group (or symmetric space) $T^N = U(1) \times U(1) \ldots \times U(1)$, that is, the N-torus. Haar measure on T^N is the product measure $\frac{dx_1}{2\pi} \cdot \ldots \cdot \frac{dx_N}{2\pi}$, which says that the x_j's are uniformly distributed independent random variables. The question of the local scaling statistics for these is well studied in the probability theory literature. It is well known [15] that the local spacings approximate a Poisson Process as $N \to \infty$. For this reason the spacings for this T^N model, as $N \to \infty$, are known as "Poisson statistics" in the physics literature. The scaled k-th consecutive spacings $\mu_k(x_1, \ldots, x_N)$ are easily shown (in the sense of (2.4) and (2.5)) to converge to $\mu_k(\text{Poisson}) := \frac{x^k e^{-x}}{k!} dx$. The limiting pair correlation $R_2(\text{Poisson})$ is simply dx on $\mathbb{R}$. The measures $\nu_1(T^N)$ converges to $e^{-x}dx$ as $N \to \infty$.

3 The Basic Conjecture

The general belief is that the local (scaled) spacing statistics of the eigenvalues of a system should be dictated by symmetry and given the symmetry type, it is universal. For the semi-classical problems as in §1.1, we must in general first scale the spectrum (or, as it sometimes is called, "unfold") before considering the spacings. For example, in the case of a compact Riemannian manifold X of dimension n, Weyl's law asserts that

$$\lambda_j \sim C_n(\text{Vol}(X)j^{2/n}, \quad \text{as} \quad j \to \infty, \qquad (3.1)$$

C_n a constant depending on n.
Hence if we set

$$\hat{\lambda}_j = \frac{\lambda_j^{n/2}}{(C_n \mathrm{Vol}(X))^{n/2}}. \tag{3.2}$$

Then $\hat{\lambda}_j \sim j$.

We examine the local spacings for the numbers $\hat{\lambda}_j$. That is, as in §1.2 set

$$\mu_k(X, N)[a, b] = \frac{\#\{j \le N \mid \hat{\lambda}_{j+k} - \hat{\lambda}_j \in [a, b]\}}{N} \tag{3.3}$$

and similarly for the other statistics. Note that $n = \dim X = 2$ is particularly pleasant in that no unfolding is necessary.

Another technical point is that if there are symmetries of the system they should be taken into account. Thus, for example, in the case of a Riemannian X, if there are discrete isometries of X one should decompose the spectrum according to these symmetries — that is, "desymmetrize".

The following is the basic conjecture of the subject. It appears by now to be a well accepted and established phenomenon in the physics literature.

Basic Conjecture

(A) (Berry-Taylor [16]). If H is completely integrable then the local spacing statistics for H in the semi-classical limit $\hbar \to 0$ or the large energy limit is Poissonian (i.e., behaves like the typical $x \in T^N$ as $N \to \infty$).

(B) (Bohigas-Giannoni-Schmit [17]). If H is chaotic then the local spacings statistics for the eigenvalues of $\hat{H}$ in the above limits follow

(i) GOE (COE) statistics if H has time reversal symmetry[2]

(ii) GUE (= CUE) statistics if H does not have time reversal symmetry.

Put another way, (A) asserts that the eigenvalues in the classically integrable case behave like random numbers while (B) asserts that the eigenvalues of a chaotic H behave like specific random matrix models, in particular they are "rigid" (an examination of the measures $\mu_1(\mathrm{GOE})$ and $\mu_1(\mathrm{GUE})$ show that their densities vanish to first and second order at $x = 0$, respectively. This signifies that for these models the eigenvalues "repel" each other).

The primary and most compelling evidence for the above are a host of numerical experiments[3], all for 2 degrees of freedom. Typical of these are the following, for which a few thousand eigenvalues can be computed. The first Hamiltonian is a billiard ball in a quarter ellipse (desymmetrized), Figure 1, $\hat{H}_E$ is the Laplacian on E with (say) Dirichlet boundary conditions. H_E is integrable (Euclid) and the spectrum is Poissonian [19]. The second is a

---

[2] That is to say, it behaves like the typical member of GOE... .

[3] and even physical experiments [18]

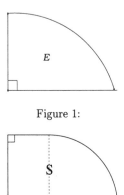

Figure 1:

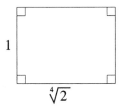

Figure 2:

billiard ball in a (quarter) Stadium S, Figure 2, considered by Bunimovich [20], who shows this classical Hamiltonian is chaotic. This H_S has time reversal symmetry (as do H_E, H_B, H_T, and H_A) and the spectrum follows GOE statistics [18,19]. Next is H_B, the billiard ball in a box with side lengths

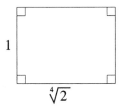

Figure 3:

shown in Figure 3. H_B is integrable and the spacings are Poissonian [16]. The next Hamiltonian H_T is the billiard motion along geodesics in a hyperbolic triangle T, Figure 4 (i.e., a triangle in the hyperbolic plane with the billiard, as always, obeying the law that angle of incidence equals angle of reflection). This is essentially the same as the geodesic flow on a surface of constant negative curvature, in particular H_T is chaotic. For the angles $\alpha = \pi/8$, $\beta = \pi/2$, $\gamma = \frac{67}{200}\pi$ and pretty much any other choice, the spectrum of $\widehat{H}_T$ (which we take to be the hyperbolic Laplacian on T with Dirichlet boundary conditions) is GOE [21].

Finally, H_A is the Hamiltonian in a triangle T as in Figure 4 except that the angles are chosen to be $\alpha = \pi/8$, $\beta = \pi/2$, $\gamma = \pi/3$. In this case very

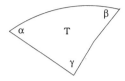

Figure 4:

surprisingly the spacings for $\widehat{H}_A$ are Poissonian! [21, 22].

All of the above, save for H_A, are phenomenological confirmations of the Basic Conjecture (we will return to this exception later). We now review the analytical results that have been established towards the Basic Conjecture.

4 Completely Integrable Hamiltonians

The basis of Conjecture A above of Berry-Tabor is their numerical experimentation with some specific examples. Recently there has been some progress in analyzing these, we discuss this next.

4.1 Particles in a box([16])

H_B of Section 3 is an example of such a Hamiltonian. These are essentially geodesic motion on a flat two dimensional torus $X_L = \mathbb{R}^2/L$. Here L is a lattice in $\mathbb{R}^2$. The angle that a geodesic makes with a given geodesic will remain constant during the geodesic flow. Thus the corresponding Hamiltonian is completely integrable. The eigenfunctions of X are $e(\langle \gamma, x \rangle)$ with $\gamma \in L^*$ the lattice dual to L. Hence the spectrum of X (or rather the Laplacian on X) is the set

$$\{4\pi^2 |\gamma|^2 : \gamma \in L^*\} \, . \tag{4.1}$$

If w_1, w_2 is a $\mathbb{Z}$-basis of L^* and $F_L(x_1, x_2) = 4\pi^2 |x_1 w_1 + x_2 w|^2$, then $\mathrm{Spec}(X_L)$ is the set of values of the quadratic form F_L at the integers $\mathbb{Z}^2$. We write $F_L(x_1, x_2) = \alpha x_1^2 + \beta x_1 x_2 + \gamma x_2^2$, $(\alpha, \beta, \gamma) \in \mathbb{R}^3$. If $(\alpha, \beta, \gamma) = \lambda(\alpha_1, \beta_1, \gamma_1)$ with $\alpha_1, \beta_1, \gamma_1 \in Q$ we say that F_L, or L, is rational. This is a singular case both classically and quantum mechanically, since the set of lengths of periodic orbits of the classical flow is highly degenerate and so is the spectrum. [16] therefore avoid these as far as Conjecture A goes, that is, they (and we) assume that L is irrational.

The pair-correlation for the spectrum of X_L is concerned with the values at integers of an irrational quadratic form of signature $(2, 2)$. In fact it is easily seen that the issue of the pair correlation being Poissonian is equivalent to a quantitative form of the classical Oppenheim Conjecture. See [23] for a review of this classical problem and its solution using the machinery of unipotent flows (Margulis [24], Ratner [25]). In [26] Eskin, Margulis, and Moses establish such a quantitative Oppenheim for irrational quadratic forms of signature $(3, 1)$. However, signature $(2, 2)$ is more subtle and the behavior depends on the Diophantine properties of the coefficients.

In [27] we were recently able to resolve this pair-correlation problem, at least for almost all X_L's, by a direct analysis. We show that for almost all X_L in the sense of Lebesgue measure on (α, β, γ), the pair correlation for $\mathrm{Spec}(X_L)$ is Poissonian. The proof involves a reduction of the problem to a delicate question about bounding integer points in $\mathbb{Z}^8$ satisfying a pair of inequalities defined by integral homogeneous forms of degree 4. Using the result about almost all X_L's one can show quite easily (see [27]) that for the topologically generic X_L (i.e., in the sense of Baire category), the consecutive spacing measures $\mu_1(X_L, N)$ do not converge as $N \to \infty$. In fact they oscillate between at least two distinct probability measures. The proof of the almost all X_L result above offers no explicit example of an X_L for which the pair-correlation is Poissonian. I learned recently from Eskin that he, Margulis, and Moses have extended the machinery used in the quantitative Oppenheim theory to include such a form as $x_1^2 + \sqrt{2}\,x_2^2 - x_3^2 - \sqrt{2}\,x_4^2$. In particular this shows that H_B of Section 3 has Poisson pair-correlation. This then is the first *explicit* example of an H for which some local spacing statistic has been shown to be what is predicted by the Basic Conjecture A.

The almost everywhere technique above generalizes. VanderKam [28] has shown that the generic, in measure, flat torus in $\mathbb{R}^4$ has Poisson pair-correlation while again for the topologically generic such X the consecutive spacings don't exist. We note that in this case one must unfold (dim $X = 4$) so that the pair-correlation is concerned with values at integers of a quartic form in eight variables.

We emphasize that the analytic results above all concern pair-correlation only.

4.2   Boxed Oscillator ([16])

The second example investigated numerically in Berry-Tabor is the Hamiltonian H_α on $T(\mathbb{R} \times S^1)$ given by

$$H_\alpha(x, \xi) = \xi_1^2 + x_1^2 + \alpha \xi_2^2 \tag{4.2}$$

This corresponds to a harmonic oscillator in one direction and a boxed (or periodic) motion in the other. Fixing $\hbar$ we have

$$\widehat{H}_\alpha = -\frac{\partial^2}{\partial x_1^2} + x_1^2 - \alpha \frac{\partial^2}{\partial x_2^2} \ . \tag{4.3}$$

Here $\alpha > 0$ is a parameter. H_α is completely integrable and the spectrum of $\widehat{H}_\alpha$ may be computed explicitly by separation of variables:

$$\text{Spec}(\widehat{H}_\alpha) = \{\lambda_{m,n} = \alpha n^2 + m \mid m \geq 1, n \in \mathbb{Z}\} \qquad (4.4)$$

To avoid obvious degeneracy we assume as we did in the last example that α is irrational. In this example we are interested in the large energy limit, $\lambda \to \infty$. The number of eigenvalues of H_α in the interval $[M, M+1]$ is approximately $\sqrt{M}$. In forming the local scaling spacing distributions such as $\mu_1(H_\alpha, M)$, we do so with the eigenvalues in $[M, M+1]$. Hence these questions reduce to the local spacing statistics for the sequence $\alpha n^2 (\text{mod} 1)$, $n \leq N$ as $N \to \infty$. While the question of the equidistribution of $\alpha n^2 (\text{mod} 1)$ and variations thereof have been studied since Weyl [29], the issue of local spacings has not been considered. (Note that $\alpha n (\text{mod} 1)$ is not random in our sense, since the consecutive spacings assume at most three values [30]).

The analogue of the pair-correlation result of (4.1) for almost all $\widehat{H}_\alpha$'s is quite easy to carry out here and was done by Rudnick and Sarnak [31]. In another work Rudnick-Zaharescu and Sarnak [32] have recently been able to go a lot further. Using Diophantine approximations to α by rationals a/q (depending on N) one can reduce the above questions to ones about the spacings between the numbers $an^2 (\text{mod} q)$, $1 \leq n \leq N$ and $q^{1/2+\varepsilon} \ll N \ll q^{1-\varepsilon}$, $\varepsilon > 0$ as $q \to \infty$ (note the ranges on N which are crucial). To calculate the k-level correlations for the last sequence one is led to estimating the number of solutions to

$$\left. \begin{array}{l} a(n_1^2 - n_2^2) \equiv b_1 (\text{mod} q) \\ \vdots \\ a(n_k^2 - n_{k+1}^2) \equiv b_k (\text{mod} q) \\ 1 \leq n_j \leq N \end{array} \right\} \qquad (4.5)$$

These equations, at least if q is prime, define a curve in affine $k + 1$ dimensional space over the finite field $\mathbb{F}_q$. One may use the Riemann Hypothesis for curves over finite fields, established by Weil [33], to estimate the number of solutions to (4.6). For N in the range $\left[q^{1-\frac{1}{2k}}, q^{1-\varepsilon}\right]$ this procedure works well and yields the answer that one would obtain if $an^2 (\text{mod} q)$ were random. This analysis leads to the Conjecture that for any irrational α, there is a subsequence $N_j \to \infty$ such that $n^2 \alpha (\text{mod} 1)$, $n \leq N_j$, has all its local spacings Poissonian. Moreover we [32] conjecture that if α is of type $\varkappa < 3$ (we say α is of type $\varkappa$ if there is a $C_\alpha > 0$ such that for all integers a, q, $(a, q) = 1$, $\left|\alpha - \frac{a}{q}\right| \geq C_\alpha q^{-\varkappa}$) then $n^2 \alpha (\text{mod} 1)$ is Poissonian along the full sequence of integers N. With the above analysis it is proven in [32] that if α is not of type 3 then, as predicted by the above Conjecture, there is a subsequence N_j of N going to infinity such that the scaled spacings of

$n^2\alpha(\mathrm{mod}\,1)$, $n \leq N_j$ are Poissonian (this is established for all the k-level correlations and hence for the consecutive spacings as well). In particular for such α (which are topologically generic but measure theoretically null) we have that $\mathrm{Spec}(\widehat{H}_\alpha)$ is fully Poissonian along a subsequence of M's that is the basic conjecture is true along a subsequence of energies. In any case the Conjectures above give precise predictions about the spacings of $\mathrm{Spec}(\widehat{H}_\alpha)$ and its dependence on the Diophantine properties of α.

4.3 Surfaces of Revolution

The above examples are special in that the spectrum could be given by an explicit formula. However in the integrable case the eigenvalues obtained by the Bohr-Sommerfeld quantization conditions [34] are sufficiently good approximations in the semiclassical limit, to yield similar working expressions. We illustrate this with X a surface of revolution of the following shape: Colin

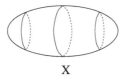

X

Figure 5:

de Verdiere [34] has shown that for such an X there are piecewise analytic functions $F_2(x_1, x_2)$ and $F_0(x_1, x_2)$, the first homogeneous of degree two and the second of degree zero, such that to an accuracy which is good enough to study scaled spacings;

$$\mathrm{Spec}(\Delta_X) = \left\{ F\left(\ell + \frac{1}{2}, k\right) : |\ell| \leq k,\ k = 1, 2, \ldots \right\} \qquad (4.6)$$

where $F = F_2 + F_0$.

In the case of a flat torus F_2 is a quadratic form, $F_0 = 0$ and the above is exact. We see that more generally the Basic Conjecture A is a question about whether the values of such an F at the integers are random? Along these lines Sinai [35] and Major [36] have shown that if F_2 is constructed from a generic (in measure) Brownian path (in particular is nowhere differentiable) then its values at the integer lattice are Poissonian. This does not apply to the above but we believe it does point to Basic Conjecture A being true in some generic sense.

Among these surfaces of revolution are some singular examples called Zoll surfaces Z [37]. These are surfaces all of whose geodesics are closed and are of the same length. They form an infinite dimensional family [37]. For these $F_2(x_1, x_2) = \frac{x_1^2}{2}$, which is degenerate and which results in the spectrum looking like a small perturbation of that of the round sphere (which is the most familiar Zoll surface). That is, the spectrum consists of clusters of $2k + 1$ eigenvalues within $O(1/k)$ of $\frac{k(k+1)}{2}$, $k \geq 1$. Thus it is clear that the consecutive spacing measures, as defined in (3.3), will approach a unit delta mass at the origin. However, given the structure of the spectrum of a Zoll surface, it is natural to form the spacing distributions out of each cluster of size $2k + 1$ and to unfold appropriately. This was done by Uribe and Zelditch [38] who determine the limiting behavior of the corresponding pair-correlation $R_2(k, Z)$ as $k \to \infty$. The result is a density on $\mathbb{R}$ determined by the geometry of the space of geodesics on Z and this density very much depends on Z. In particular it is non-universal.

This concludes our review of Case A of the Basic Conjecture. Examples (4.1) and (4.2) show that the problem is very subtle in its dependence on parameters. The Basic Conjecture A is true at best in some generic sense. The examples indicate that generic should be in the sense of almost-all with respect to a measure on the space of Hamiltonians under consideration. A precise formulation, however, is missing at this time.

5 Chaotic Case

An analytic attack on the Basic Conjecture B for chaotic H's is much more difficult. There are no examples with explicitly computable spectra. Moreover there is no semi-classical approximation of the Bohr-Sommerfeld type (this being the reason that Bohr was unsuccessful at quantizing anything but completely integrable H's). The only rigorous tools available are the trace formula (Selberg [39], Gutzwiller [40], Balian-Bloch [41], and Duistermaat-Guillemin [42]). These give expressions for sums over the spectrum of $\hat{H}$ in terms of sums over the periodic orbits of the classical flow. The simplest case and the only one for which there is an exact formula is that of Selberg which in the case of X being a hyperbolic surface (i.e., constant curvature $K \equiv -1$) of genus $b \geq 2$ reads as follows [39]:

For any $g \in C_0^\infty(\mathbb{R})$, $g(x) = g(-x)$ and $h(t) = \hat{g}(t)$ its Fourier transform

$$\sum_{j=0}^{\infty} h(t_j) = (b-1) \int_{-\infty}^{\infty} th(t) \tanh(\pi t) dt + \sum_{p \in P} \sum_{k=1}^{\infty} \frac{\ell(p)}{e^{k\ell(p)/2} - e^{-k\ell(p)/2}} g(k\ell(p))$$

(5.1)

where $\lambda_j = t_j^2 + \frac{1}{4}$ are the eigenvalues of Δ_X, and P is the set of prime closed geodesics on X and for $p \in P$, $\ell(p)$ is its length.

For the analysis of sums over the spectrum λ_j in intervals $[\lambda, \lambda + R]$ where $R \geq \sqrt{\lambda}$, the above trace formula is very powerful since on the right hand side there are essentially no contributions from the periodic orbits. However for $R = \lambda^\alpha$ with $\alpha < \frac{1}{2}$, the right hand side involves sums over exponentially many periodic orbits and is very difficult to analyze. In particular one can express the pair correlation for spec(Δ_X) in terms of such sums over periodic orbits. However the problem of establishing cancellations over these long sums over periodic orbits is beyond the range of present technology. Consequently there are no proven results concerning the local spacings for the spectrum of such X's.

One way to proceed is to impose some assumptions about the statistical nature of the long periods $l(p)$, $p \in P$. While it appears difficult to verify such assumptions (for any given X or even on average X's) it is interesting to see what these lead to. The results of Berry [43] and the more recent deeper investigation of Bogomolny and Keating [44] lead to results about the local spacing for spec(Δ_X) which are consistent with GOE (note that geodesic flow always has time reversal symmetry). However it remains problematic, even with these assumptions, to recover the GOE pair correlation, (2.10).

One aspect that is clarified by an analysis using the trace formula is that if the periodic orbit spectrum is highly degenerate then the local spacing will not be GOE. Indeed if X is a hyperbolic surface as above then X may be realized as $\Gamma \backslash \mathbb{H}^2$ where $\mathbb{H}^2$ is the hyperbolic plane and Γ is a discrete co-compact subgroup of $SL_2(R)$. Among such Γ's are such groups as $SL_2(Z)$ (which has finite volume but not compact quotient) which are defined through integers and are called arithmetic (see Borel [45] for definition). There are 85 hyperbolic triangles T in H^2 for which the reflections in their sides generate such a discrete arithmetic Γ (Takeuchi [46]). Example H_A of Section 3 is one of these. If $X = X_\Gamma$ is arithmetic then the length spectrum is very highly degenerate [LS]. For these one shows that:

$$\sum_{l(p)\leq V,\ \text{distinct } l(p)'s, p \in P} 1 = O(e^{V/2}) \qquad (5.2)$$

as $V \to \infty$.

On the other hand it follows easily from (5.1) that for any hyperbolic X

$$\sum_{l(p)\leq V, p \in P} 1 \sim \frac{e^V}{V} \text{as } V \to \infty \qquad (5.3)$$

Hence in the arithmetic case the average degeneracy of a length of a closed geodesic of length l is about $e^{l/2}$. As was first observed in [47] and [48], this degeneracy manifests itself in the local spacing statistics of the spectrum of Δ_{X_Γ} being non GOE (actually, Selberg had noticed long before that this degeneracy forces the remainder term in Weyl's law to be very large - see Hejhal [49]). Numerical experiments with a large subset (H_A being one such) of the 85 arithmetic triangles show that they have Poisson spacing statistics! Various levels of analytic explanations of this phenomenon have been given using the trace formula (5.1) (see [22, 47, 48, 50, 51]). We note these triangles have no extra symmetries. However there is a large family of geometrically defined operators on $L^2(X_\Gamma)$ known as Hecke operators, which commute with Δ_{x_Γ} [3]. Moreover there are many such Hecke operators iff Γ is arithmetic [52].

A completely different approach to the basic conjecture B has been put forth by Agam, Altshuler, Andreev and Simons [53]. Their ideas come from the theory of "disordered systems" such as the spectral properties of $H = \Delta + V$ in $L^2(R^3)$, where V is a random potential. For these, when averaging over the potential V, Efetov [54] has shown how one directly arrives at the random matrix correlations in the extended state regime, by using a functional integral representation for the Greens function of $\hat{H}$. These integrals are analyzed using supersymmetric methods and in particular non-linear σ-models [54]. [53] adapt these methods to the case of $\hat{H}$ with H chaotic. Unlike the case above where averages over V are performed, they attempt to deal with an individual Hamiltonian H. At present their conclusions appear to be the strong - for example they assert that for a triangle such as H_A of Section 3 (which satisfies their assumptions about chaotic H and about location of poles of the corresponding Ruelle zeta function) the spacings are GOE. Nevertheless their approach is interesting and in particular its strength may lie in studying ensembles of H's with the idea of establishing that the GOE is valid for the random H (like the case of random V) or for a typical H in the sense of measure, from the ensemble. In fact, very recently Zirnbauer [55] has pursued this possibility for the quantization of chaotic symplectic transformations and his approach looks promising. In passing we note that there is a growing literature concerning the Basic Conjecture for symplectomorphisms instead of Hamiltonians, see Zelditch [4] for review of mathematical results in this direction.

We conclude Lecture 1 with some comments about the Basic Conjecture B. Our understanding of this case is a lot poorer than that for A. In part, this is due to the lack of examples that can be studied analytically. The known exceptions to GOE, that is the arithmetic X's, are sparse. In particular consider the analogue of the spaces of flat tori, which are the moduli spaces $\mathcal{M}_b$ of hyperbolic surfaces of genus b. $\mathcal{M}_b$ is a $3b - 3$ complex manifold [56] and the set of arithmetic X's in a given $\mathcal{M}_b$ is finite [45]. One certainly expects that, measure-wise, the generic X in $\mathcal{M}_b$ satisfies GOE. With our present minimal understanding it is even possible that Basic Conjecture B is valid for all but finitely many X's in $\mathcal{M}_b$ though this seems to me to be unlikely.

In the second lecture we examine similar questions for the spacing and densities of zeros of zeta functions. Here much more progress on the analytic side is possible and a solid picture has emerged.

References

[1] V. Arnold and A. Avez, *Ergodic problems of Mechanics* (W. A. Benjamin, New York, 1968).

[2] Y. Sinai, *Math. Notes.* **18**, (1976), princeton University Press, Princeton.

[3] P. Sarnak, *The Schur Lectures*, Vol. 8 of *Israel Math. Conf. Proc.* (AMS, Tel Aviv 1992, 1995), pp. 183–236.

[4] S. Zelditch, Quantum dynamics from a semi-classical point of view, 1997, preprint.

[5] N. Hurt, *MIA* **397**, (1997), kluwe Academic Publishers, Boston.

[6] R. Heller, *Phys. Rev. Lett.* **53**, 1515 (1984).

[7] E. Wigner, *SIAM Review* **9**, 1 (1967).

[8] M. Gaudin, *Nuclear Phys.* **25**, 447 (1961).

[9] M. L. Mehta, *Random Matrix Theory* (Academic Press, N.Y., 1991).

[10] F. Dyson, *J. Math. Phys.* **3**, 166 (1962).

[11] A. Altland and M. Zirnbauer, *cond-mat/9602137* (1996).

[12] N. Katz and P. Sarnak, *Random matrices, Frobenius eigenvalues and monodromy* (AMS, Rhode Island, 1998), to appear.

[13] N. Katz and P. Sarnak, Zeros of zeta functions, their spacings and their spectral nature, 1997, preprint.

[14] S. Helgason, *Differential geometry, Lie Groups, and symmetric spaces* (Academic Press, New York, 1978).

[15] W. Feller, *An introduction to probability theory and its applications. Vol. II* (John Wiley & Sons Inc., New York, 1966).

[16] M. Berry and M. Tabor, *Proc. Roy. Soc. London A* **356**, 375 (1977).

[17] O. Bohigas and M.Giannoni, *Phys. Rev. Lett.* **52**, 1 (1984).

[18] H. D. Gräf et al, *Phys. Rev. Lett.* **69**, 1296 (1992).

[19] O. Bohigas, M. Giannoni, and C. Schmit, *J. Physique Lett.* **45**, (1984).

[20] L. Bunimovich, *Functional Analysis and Applications* **8**, 254 (1974).

[21] C. Schmit, *Chaos and Quantum Physics (in Les Houches, conference)* (North Holland, Amsterdam, 1991), pp. 333–369.

[22] R. Aurich, E. Bogomolny, and F. Steiner, *Physica D* **48**, 76 (1991).

[23] A. Borel, *BAMS* **32**, 184 (1995).

[24] G. Margulis, *C. R. Acad. Sci. Paris SerI Math.* **304**, 247 (1987).

[25] M. Ratner, *Ann. Math.* **134**, 545 (1991).

[26] A. Eskin, G. Margulis, and S. Moses, *Electronic Res. Announcements AMS* **1**, 124 (1995).

[27] P. Sarnak, Values at integers of Binary Quadratic Forms, to appear in a volume in memory of C. Herz, edited by S. Drury.

[28] J. Vanderkam, Pair correlation of four-dimensional flat tori, 1997, preprint.

[29] H. Weyl, *MAth. Ann.* **77**, 313 (1916).

[30] S. Swierczowski, *Fund. Math.* **46**, 187 (1958).

[31] Z. Rudnick and P. Sarnak, *Duke Math. J.* **81**, 269 (1996).

[32] Z. Rudnick, P. Sarnak, and A. Zahareschu, in preparation.

[33] A. Weil, *Proc. Nat. Acad. Sci. U.S.A.* **27**, 345 (1941).

[34] Y. Colin de Verdiere, *Invent. Math.* **43**, 15 (1977).

[35] Y. Sinai, *Dynamical Systems and Stat. Mech., Adv. Soviet Math.* **3**, 199 (AMS 1991).

[36] P. Major, *Probab. Theory Related Fields* **92**, 423 (1992).

[37] A. Besse, *Manifolds all of whose geodesics are closed* (Springer-Verlag, N.Y., 1978).

[38] A. Uribe and S. Zelditch, *Com. Math. Phys.* **154**, 313 (1993).

[39] A. Selberg, *J. Ind. Math. Soc.* **20**, 47 (1956).

[40] M. Gutzwiller, *J. Math. Phys.* **8**, 1979 (1967).

[41] B. Balian and C. Bloch, *Ann. Phys.* **69**, 76 .

[42] H. Duistermaat and V. Guillemin, *Invent. Math.* **29**, 39 (1975).

[43] M. Berry, *Proc. Royal. Soc. London* **A 400**, 229 (1985).

[44] E. Bogomolny and J. Keating, *Phys. Rev. Lett.* **77**, 1472 (1996).

[45] A. Borel, *Ann. Sc. Norm. Super. Pisa Sci(4)* **8**, 1 (1981).

[46] K. Takeuchi, *J. Fac. Sci. Univ. Tokyo* **24**, 201 (1977).

[47] E. Bogomolny, B. Georgeot, M. Giannoni, and C. Schmit, *Phys. Rev. Lett.* **69**, 1477 (1992).

[48] F. S. J. Bolte, G. Steil, *Phys. Rev. Lett.* **69**, 288 (1992).

[49] D. Hejhal, *S.L.N.* **548**, (1976), springer, N.Y.

[50] W. Luo and P. Sarnak, *Com. Math. Phys.* **161**, 423 (1994).

[51] E. Bogomolny, F. Leyoraz, and C. Schmit, *Comm. Math. Phys.* **176**, 577 (1996).

[52] G. Margulis, in *Proc. Int. Congr. Math.* (Canadian Math. Congress, Vancouver, 1975), p. 21.

[53] O. Agam, B. Simons, A. Andreev, and B. Altshuler, *Nucl. Phys. B* **482**, 536 (1996).

[54] K. B. Efetov, *Advances in Physics* **32**, 53 (1983).

[55] S. Zirnbauer, The color-flavor transformation and a new approach to quantum chaotic maps, 1997, preprint.

[56] A. Weil, *Collected Works* (Springer, N.Y., 1980), Vol. II, pp. 381–391.

Quantum Chaos, Symmetry and Zeta Functions
Lecture II: Zeta Functions

Peter Sarnak

Department of Mathematics
Princeton University
Princeton, NJ 08544
sarnak@math.princeton.edu

1 The Zeta Function

In the first lecture we discussed the role played by symmetry in local spacing statistics for quantizations of classical Hamiltonians. In this lecture we discuss the role of symmetry in the local spacing distribution between zeros of zeta functions.

We begin with Riemann Zeta Function $\zeta(s)$ and some phenomenology associated with it:

$$\zeta(s) = \sum_{n=1}^{\infty} n^{-s} = \prod_{p} (1 - p^{-s})^{-1} \tag{1.1}$$

which converges for $\mathrm{Re}(s) > 1$. $\zeta(s)$ has an analytic continuation and functional equation [1]:

$$\xi(s) := \pi^{-s/2}\Gamma(s/2)\zeta(s) = \xi(1 - s) \tag{1.2}$$

$\xi(s)$ has simple poles at $s = 0$ and $s = 1$ and is otherwise analytic. Write the zeros ρ_j of $\xi(s)$ as:

$$\rho_j = \frac{1}{2} + i\gamma_j. \tag{1.3}$$

From (1.1) it is clear that $|\Im(\gamma_j)| \leq \frac{1}{2}$. The well known Riemann Hypothesis "R-H" asserts that $\gamma_j \in \mathbb{R}$. For the following questions of local spacings, let's assume RH (in numerical experiments that are quoted this has been verified for the zeros examined). Order the zeros

$$\ldots \leq \gamma_{-3} \leq \gamma_{-2} \leq \gamma_{-1} \leq \gamma_1 \leq \gamma_2 \ldots \tag{1.4}$$

here $\gamma_{-j} = -\gamma_j, j = 1, 2, \ldots$. It is well known [2] that:

$$|\{j \geq 1 | \gamma_j \leq T\}| \sim \frac{T \log T}{2\pi} \tag{1.5}$$

as $T \to \infty$. Hence we form the local spacings by unfolding:

$$\hat{\gamma}_j := \frac{\gamma_j \log \gamma_j}{2\pi} \tag{1.6}$$

The $\hat{\gamma}_j$'s have mean spacing one.

During the years 1980-1997 Odlyzko [3] has made an extensive and profound numerical study of these zeros and in particular of the local spacings of $\hat{\gamma}_j$. He found that these obey the GUE model perfectly. For example in Figure 1.14 of [4] of Lecture 1, the consecutive spacings for the $7\mathrm{x}10^7$ zeros beyond the 10^{20}-th zero are plotted against the density $\mu_1(\mathrm{GUE})$. At the phenomenological level this is perhaps the most striking discovery about the zeta function since Riemann. The big question is why is this so and what does it tell us about the nature (e.g. spectral) of the zeros. Also what is the symmetry which is responsible for this GUE or type II symmetric space statistics (cf. (2.4) of Lecture I).

Odlyzko's computations were inspired by the 1974 discovery of Montgomery [5] that the pair-correlation is, at least for a restricted class of test functions, equal to the GUE pair-correlation. Precisely he proves that as $N \to \infty$,

$$\frac{1}{N} \sum_{\substack{1 \leq j \neq k \leq N}} \phi(\hat{\gamma}_j - \hat{\gamma}_k) \to \int_{-\infty}^{\infty} \phi(x) R_2(GUE)(x) dx \tag{1.7}$$

for any $\phi \in \mathcal{S}(\mathbb{R})$ for which the support of $\hat{\phi}$ is contained in $(-1, 1)$ where $\hat{\phi}(\xi) = \int_{-\infty}^{\infty} e^{-2\pi i x \xi} \phi(x) dx$. Note that $\hat{R}_2(\xi)$ (see (2.8) of Lecture I) changes its analytic character at $\xi = \pm 1$ and indeed the terms contributing to (1.7) come from the "diagonal" [5, 6]. Extending (1.7) to $\hat{\phi}$'s whose support is not contained in [-1,1] involves new non-diagonal contributions and this has never been achieved (see Goldston Montgomery [7] for an equivalence). Note that this already (albeit assuming RH) goes far beyond anything that one can establish for the pair-correlation for the t_j's in Lecture I. The reason is that the unfolding of γ_j is $\gamma_j \log \gamma_j$ while for t_j it is t_j^2. This has the effect on the right hand side of the analogue of (5.1) of Lecture I for $\zeta(s)$ (known as the explicit formula see [6]) of facing $\log p \leq \log T$ terms when support $\hat{\phi} \subset (-1, 1)$, while for the t_j case we always have e^T terms (this has been referred to as the exponential proliferation of periodic orbits in the latter case).

More recently Hejhal [8] used Montgomery's method to establish that the triple correlation is the GUE triple correlation as computed in Dyson [9]. Rudnick and Sarnak [6] by a somewhat different method (which in fact does not require RH) establish that all the $n \geq 2$ correlations are as predicted by GUE. All of these results are restricted as above, that is they are proven in the range of the Fourier transforms where only the "diagonal" contributions constitute the main term. An interesting heuristic derivation of the n-level correlations without any restrictions on the Fourier transforms has been given by Bogomolny and Keating [4].

The zeta function is but the first of the zoo of L-functions for which similar questions can be asked. There are the Dirichlet L-functions $L(s, \chi)$ defined as follows: $q \geq 1, \chi : (\mathbb{Z}/q\mathbb{Z})^* \to S^1$ is a character and extend χ to $\mathbb{Z}$ by setting χ to be periodic of period q and $\chi(m) = 0$ if $(m, q) \neq 1$. Then

$$L(s, \chi) = \sum_{n=1}^{\infty} \frac{\chi(n)}{n^s} = \prod_p (1 - \chi(p)p^{-s})^{-1}. \tag{1.8}$$

The analogue of (1.2), that is the analytic continuation and functional equation are known for these. Even more generally we have for each automorphic cusp form f on $GL_m/\mathbb{Q}$ [10] an L-function $L(s, f)$, which satisfies similar properties [11]. A classical concrete form on $GL_2/\mathbb{Q}$ is the form $\Delta(q)$ [12],

$$\Delta(q) := q \prod_{m=1}^{\infty} (1 - q^m)^{24} = \sum_{n=1}^{\infty} \tau(n)q^n \tag{1.9}$$

$$L(s, \Delta) = \sum_{n=1}^{\infty} \frac{\tau(n)}{n^{11/2}} n^{-s} = \prod_p (1 - \frac{\tau(p)}{p^{11/2}} p^{-s} + p^{-2s})^{-1}. \tag{1.10}$$

In general the L-function of an automorphic form on $GL_m/\mathbb{Q}$ is an Euler product of local factors of degree m in p^{-s}. In all these cases an R-H for $L(s, f)$ is expected to hold.

The results of Rudnick-Sarnak [6] were carried out in this context and they show that the $n \geq 2$ correlations (for ϕ's even more restricted as m grows) are universally the GUE ones! Moreover at the numerical level Rumely [13] has checked that the zeros of Dirichlet L-functions satisfy GUE statistics and Rubinstein [14] has checked various GL_2 L-functions and finds that they all satisfy GUE local spacing statistics. We call this phenomenon, that the high zeros of any L-function $L(s, f)$, f a cusp form on $GL_m/\mathbb{Q}$ obey GUE spacing laws, the "Montgomery - Odlyzko Law".

2 Function Fields

One can get much insight into the source of the Montgomery-Odlyzko Law by considering its function field analogue. The function field analogue of $\zeta(s)$ is due to Artin [15]. If k is a finite extension of the field $\mathbb{F}_q$ of rational functions with coefficients in the finite field $\mathbb{F}_q$, its zeta function $\zeta(k, T)$ is defined to be

$$\zeta(k, T) = \prod_v (1 - T^{\deg(v)})^{-1} \qquad (2.1)$$

where the product is over all the places v of k [15]. One can also think of $\zeta(k, T)$ as the zeta function of a nonsingular curve $C/\mathbb{F}_q$ whose field of functions is k. This geometric point of view is very powerful. For example the Riemann-Roch Theorem on the curve plays the role of Poisson-Summation in (1.2) and it yields [16]

$$\zeta(k, T) = \frac{P(k, T)}{(1 - T)(1 - qT)} \qquad (2.2)$$

where P is a polynomial of degree $2g$, g being the genus of k, as well as a functional equation for $\zeta(k, T)$ when T is replaced by $1/qT$. The Riemann Hypothesis for $\zeta(k, T)$ asserts that all the zeros of P lie on the circle $|T| = q^{-1/2}$. This was established by Weil [17]. A key point in this proof is the interpretation of the zeros of $\zeta(k, T)$ as the reciprocals of the eigenvalues of Frobenius (which is the operation of raising the coordinates of points on the corresponding curve C to the power q) acting on the first cohomology groups of the curve C [17].

Turning to the distribution of the zeros of such a zeta function in (2.1), we write its zeros as:

$$\zeta_j = e^{i\theta_j} q^{-1/2}, j = 1, \ldots, 2g. \qquad (2.3)$$

Form the local spacing measures as in (2.1) and (2.2) of Lecture I and denote them by $\mu_k(C/\mathbb{F}_q)$. For a fixed $\zeta(C/\mathbb{F}_q, T)$ there are only $2g(C)$ zeros and so we cannot have a spacing law. We therefore let the genus $g = g(C)$ go to infinity. However this alone will not allow one to deduce a unique limiting law since there are curves $C/\mathbb{F}_q$ of large genus which have a large number of symmetries and for which the local spacings are Poissonian, see [18]. In Katz-Sarnak [18] we therefore consider the typical curve of large genus g and over a large field $\mathbb{F}_q$. We show [18] that as q and g go to infinity the local spacings follow the GUE model, that is the Montgomery-Odlyzko Law is valid for these zetas. Precisely if $\mathcal{M}_g(\mathbb{F}_q)$ denotes the set of isomorphism classes of curves of genus g over $\mathbb{F}_q$, then $k \geq 1$:

$$\lim_{g \to \infty} \lim_{q \to \infty} \frac{1}{|\mathcal{M}_q(\mathbb{F}_q)|} \sum_{C \in M_q(\mathbb{F}_q)} D\left(\mu_k(C/\mathbb{F}_q), \mu_k(GUE)\right) = 0 \qquad (2.4)$$

Note that the double limit must be carried out in the order indicated. The key ingredients in the proof of 2.4 are:

- The monodromy group of the family $\mathcal{M}_g$ (or more accurately a closely related family) [18] of curves of genus g, which arises through the representation of the fundamental group of the family on the first cohomology group at a base curve. The first homology group comes with an intersection pairing for cycles and the symplectic pairing is preserved by the monodromy. In fact the monodromy turns out to be the full symplectic group $Sp(2g)$ and this is the key point.

- The Equipartition Theorem of Deligne [19,20] for the Frobenii for the family, in the monodromy, as $q \to \infty$.

- The Law of Large Numbers (2.5) of Lecture I for the scaling limits of $USp(2g)$ as $g \to \infty$.

Thus in the function field, the source of the GUE is clearly identified. In part it is due to the universality of the local statistics for type II symmetric spaces (2.4) of Lecture I. Also there is a symmetry behind the GUE law — it comes from the scaling limits of the monodromies of the family. We again see that it is more reasonable, at least to begin with, to examine these local spacing statistics for families. In this function field case, at least if the monodromies of the families and their scaling limits can be computed — then one has a complete understanding (at least on letting $q \to \infty$ as is done in (2.4)).

3   Families in the Global case

We return to global zeta or L-functions, that is $L(s, f)$ where f is an automorphic cusp form of $GL_m/\mathbb{Q}$, and consider families $\mathcal{F}$, of such. We do not offer a precise definition of what is meant by a family in this case, but rather (since this is all that we have at present) we give numerous examples of families. The set up is such that each $f \in \mathcal{F}$ has a "conductor" $c_f \in (0, \infty)$ (they are given explicitly in the examples below). For X a real parameter, we assume that the sets $\mathcal{F}_X = \{f \in \mathcal{F} | c_f \leq X\}$ are finite and that the asymptotics of $|\mathcal{F}_X|$ as $X \to \infty$, are known. The scaling statistics which we consider are the distributions of zeros of $L(s, f)$ near $s = 1/2$, as f varies over $\mathcal{F}$ ordered by conductor. That is we examine the numbers:

$$\frac{\gamma_f^{(j)} \log c_f}{2\pi}, 0 \leq \gamma_f^{(1)} \leq \gamma_f^{(2)} \ldots, \tag{3.1}$$

where $\frac{1}{2} + i\gamma_f^{(j)}$ are the nontrivial zeros of $L(s, f)$. That the scaling by $\frac{\log c_f}{2\pi}$ is appropriate will become clear from the results below. To measure these distributions we form for $\mathcal{F}$ the analogues of the measures ν_k (e.g. (2.11) of Lecture I) and the densities D_1 (e.g. (2.12) of Lecture I) as follows:

$$\nu_j(X, \mathcal{F})[a, b] = \frac{1}{|\mathcal{F}_X|} \# \left\{ f \in \mathcal{F} : c_f \leq X, \frac{\gamma_f^{(j)} \log c_f}{2\pi} \in [a, b] \right\} \tag{3.2}$$

and for $\phi \in \mathcal{S}(\mathcal{R})$ a test function set:

$$D(f, \phi) = \sum_{\gamma_f} \phi(\frac{\gamma_f \log c_f}{2\pi}) \tag{3.3}$$

and

$$W(X, \mathcal{F}, \phi) = \frac{1}{|\mathcal{F}_X|} \sum_{c_f \leq X} D(f, \phi). \tag{3.4}$$

Thus $\nu_j(X, \mathcal{F})$ measures the distribution, as f varies over $\mathcal{F}$, of the j-th lowest zero of $L(s, f)$ normalized as in (3.1), while W measures the density the zeros which are within $O(1/\log c_f)$ of $s = 1/2$. One hopes that as $X \to \infty$ the measures $\nu_j(X, \mathcal{F})$ converge to measures $\nu_j(\mathcal{F})$ and the densities converge to $\int_{-\infty}^{\infty} \phi(x) W(\mathcal{F})(x) \, dx$, for a suitable density $W(\mathcal{F})(x)$. Indeed for the function field analogue of the above with various families $\mathcal{F}$, this is proven in [18] using the same methods mentioned in Section 2. For these cases the limiting measures $\nu_j(\mathcal{F})$ and the density $W(\mathcal{F})$ are determined by the "symmetry" $G(\mathcal{F})$ which is the scaling limit of the monodromy groups. They are determined (when G is one of the classical families) by (2.14) and

(2.15) of Lecture I. We now list some examples of families of such $\mathcal{F}$'s for which some results along these line have been established. In all cases we will assume RH for all L-functions (at the cost of restricting the test functions further one can remove this assumption).

I: The family $\mathcal{F}$ of Dirichlet L-functions $L(s, \chi)$ where χ is a primitive quadratic character (that is $\chi^2 = 1$), mod q. The conductor c_χ is equal to q.

- From the function field analogue we expect that $G(\mathcal{F}) = Sp(\infty)$, see [21].

- [21] (see also Ozluk-Snyder [22]) $W(X, \mathcal{F}, \phi) \to \int_{-\infty}^{\infty} \phi(x)\omega_1(Sp, x)dx$ as $X \to \infty$ for $\phi \in \mathcal{S}(\mathcal{R})$ with support $\hat{\phi} \subset (-2, 2)$. Here $\omega_1(Sp, x)$ is given in (2.16) of Lecture 1.

- Rubinstein [14] has investigated numerically the distributions of $\nu_j(\mathcal{F}, X), j = 1, 2$ and $W(X, \mathcal{F})$ for $X \approx 10^{12}$ and finds an excellent fit with the $Sp(\infty)$ predictions.

- The first to numerically compute zeros $L(s, \chi)$ in this family for moderate sized q appears to be Hazelgrove. He found that the zeros 'repel' the point $s = 1/2$ and this is sometimes called Hazelgrove's phenomenon. Now the density of $\nu_1(Sp)$ vanishes to second order at 0 (see [18]) and this is unique to the Sp symmetry! So this Hazelgrove phenomenon is a manifestation of the symplectic symmetry.

II: The family $\mathcal{F}$ of quadratic $L(s, \Delta \otimes \chi)$ of the GL_2 cusp form Δ of weight 12 for $\Gamma = SL_2(\mathbb{Z})$, see (1.9) above.

$$L(s, \Delta \otimes \chi) := \sum_{n=1}^{\infty} \frac{\tau(n)}{n^{11/2}} \chi(n) n^{-s}. \tag{3.5}$$

The conductor $c_{\Delta \otimes \chi}$ is q^2 (where χ has conductor q). In this family half of the L-functions have even functional equations and half, odd functional equation, according to the sign of the "epsilon factor" $\epsilon_{\Delta \otimes \chi}$. We let $\mathcal{F}^{\pm}$ be the corresponding subfamilies.

- From the function field analogue we expect that $G(\mathcal{F}) = O(\infty)$. In particular $G(\mathcal{F})$ corresponds to the scaling limit through $O^+(even) = SO(even)$ or $O^-(odd)$ and $G(\mathcal{F}^-)$ to $O^+(odd) = SO(odd)$ or $O^-(even)$, see [21].

- ([21]) $W(X, \mathcal{F}^+, \phi) \to \int_{-\infty}^{\infty} \phi(x)\omega_1(SO(even), x)dx$, $W(X, \mathcal{F}^-, \phi) \to \int_{-\infty}^{\infty} \phi(x)\omega_1(SO(odd), x)dx$ for ϕ with support $\hat{\phi} \subset (-1, 1)$. The explicit densities $\omega_1(SO(even))$ of $\omega_1(SO(odd))$ are given in (2.16) of Lecture I.

- Numerical experimentations by Rubinstein [14] with $\nu_j(X, \mathcal{F}^\pm), j = 1, 2$ and $W(X, \mathcal{F}^\pm)$ with $X \approx 10^6$, agree well with the $O(\infty)$ predictions.

III: The family $\mathcal{F}$ of holomorphic (Hecke-eigen)-cusp forms of even integral weight k on $SL_2(\mathbb{Z}) \setminus \mathbb{H}^2$ (see [12]) as $k \to \infty$. For $f \in \mathcal{F}$, $L(s, f)$ is its L-function and its conductor is $C_f = k^2$. As in the last example half of these $L(s, f)$'s have even functional equations and half odd. In fact the sign ϵ_f is 1 if $k \equiv 0(4)$ and -1 if $k \equiv 2(4)$. Let $\mathcal{F}^\pm$ be the corresponding subfamilies.

- We expect, since the f's are generic GL_2 forms, that $G(\mathcal{F}) = O(\infty)$.

- Iwaniec-Luo-Sarnak [23] show that $W(X, \mathcal{F}^+, \phi) \to \int_{-\infty}^{\infty} \phi(x) \times \times \omega_1(SO(even), x)dx$ and $W(X, \mathcal{F}^-, \phi) \to \int_{-\infty}^{\infty} \phi(x)\omega_1(SO(odd), x)dx$, for $\hat{\phi}$ supported in (-2,2).

IV: The family $\mathcal{F}$ of holomorphic new-forms of a fixed even integral weight $k \geq 2$ for $\Gamma_0(N) \setminus \mathbb{H}$ [24][1], with $N \to \infty$. We assume that the central character of f is trivial (i.e. trivial Nebentypus) and for simplicity we also assume that N is prime. This time we average over smaller families - that is over all f's above on $\Gamma_0(N)$, with $N \to \infty$. The conductor c_f is N and, as in the last two examples, approximately half of the signs ϵ_f are +1 and half -1. Let $H_k(N)$ denote the set of forms as above and $H_k^\pm(N)$ the subsets whose corresponding $\epsilon_f = \pm 1$.

- As in the last family we expect that $G(\mathcal{F}) = O(\infty)$.

- Iwaniec-Luo-Sarnak [23] prove that as $N \to \infty$

$$\frac{1}{|H_k^+(N)|} \sum_{f \in H_K^+(N)} D(\phi, f) \to \int_{-\infty}^{\infty} \phi(x)\omega_1(SO(even), x)dx.$$

$$\frac{1}{|H_k^-(N)|} \sum_{f \in H_K^-(N)} D(\phi, f) \to \int_{-\infty}^{\infty} \phi(x)\omega_1(SO(odd), x)dx.$$

for any $\phi \in \mathcal{S}(\mathbb{R})$ support $\hat{\phi} \subset (-2, 2)$.

V: The family of symmetric square L-functions, $L(S, \vee^2 f)$ (see [25]), where F is in family III. There are Euler products of degree three and, by a Theorem of Gelbart and Jacquet [10], they are L-functions of selfdual cusp froms on GL_3. The conductor $c_{\vee^2 f}$ is k^2. The sign of the functional equation $\epsilon_{\vee^2 f}$ is always equal to 1.

---

[1]Here

$$\Gamma_0(N) = \left\{ \begin{pmatrix} a & b \\ c & d \end{pmatrix} \in SL_2(\mathbb{Z}) : N|c \right\}$$

- Being generic selfdual forms on GL_3 we expect $G(\mathcal{F}) = Sp(\infty)$.

- In [23] it is proven that

$$W(X, \mathcal{F}, \phi) \to \int_{-\infty}^{\infty} \phi(x)\omega_1(Sp, x)dx$$

as $X \to \infty$ for any $\phi \in \mathcal{S}(\mathbb{R})$ with support $\hat{\phi} \subset (-\frac{4}{3}, \frac{4}{3})$.

Remarks:

1. All of the above results confirm, to the extent that they apply, the predictions of the claimed symmetry $G(\mathcal{F})$. The Conjecture that the density sums $W(X, \mathcal{F}, \phi)$ converge to the claimed density without any restrictions on $\hat{\phi}$, will be called the Density Conjecture for the family $\mathcal{F}$.

2. The proofs of the results about the densities all proceed by expressing $D(f, \phi)$ via the explicit formula in terms of sums involving the Hecke eigenvalues of f. What is then needed are techniques for averaging the latter over $f \in \mathcal{F}$. For the families III, IV and V we use heavily the tools developed in Iwaniec-Sarnak [26] (see below) for dealing with these averages.

3. With the exception of II, all the results allow for the support of $\hat{\phi}$ to be larger than [-1,1]. This is rather significant since $\hat{\omega}_1(Sp)(\xi)$, $\hat{\omega}_1(SO(even))(\xi)$ and $\hat{\omega}_1(SO(odd))(\xi)$ are all discontinious at $\xi = \pm 1$. This signals that new terms ("nondiagonal")enter into the main terms of the asymptotics as soon as support $\hat{\phi}(\xi)$ is larger than [-1,1]. Thus what is shown here goes beyond anything established for the correlations of high zeros (see the discussion following (1.7)), or for that matter the diagonal analysis of Berry [27] (see the discussion after (5.1) in Lecture I in the analogous analysis with the trace formula). For the families III, IV and V these new non-diagonal terms arise out of a far reaching analysis with Kloosterman sums (see [26] and [28] for related issues). That these fundamentally new nondiagonal contributions yield the conjectured $G(\mathcal{F})$ answers is very pleasing evidence for the conjectures.

4 Applications

The interest in the zeros of L-functions lies in their fundamental influence on arithmetical problems. For example the question of vanishing of an L-function at special points on the critical line arises in the Birch and

Swinnerton-Dyer Conjectures [29, 30], in the Shimura correspondence (see [31]) and in spectral deformation theory (Phillips-Sarnak [32]). The distribution of zeros near $s = 1/2$ (that is the central value) discussed in Section 2.3 has immediate application to nonvanishing at this point. By varying the test function ϕ in the Density Conjecture for any of the above families $\mathcal{F}$, together with the fact that $W(\mathcal{F})$ does not give positive mass to the point zero, implies (assuming the Density Conjecture) that as $X \to \infty$,

$$\frac{\#\{f \in \mathcal{F}|c_f \leq X, \epsilon_f = 1, L(\frac{1}{2}, f) \neq 0\}}{\#\{f \in \mathcal{F}|c_f \leq X, \epsilon_f = 1\}} \to 1, \tag{4.1}$$

and

$$\frac{\#\{f \in \mathcal{F}|c_f \leq X, \epsilon_f = -1, L'(\frac{1}{2}, f) \neq 0\}}{\#\{f \in \mathcal{F}|c_f \leq X, \epsilon_f = -1\}} \to 1. \tag{4.2}$$

The results of the last Section are approximations to the density Conjecture and give corresponding approximations to (4.1) and (4.2). We illustrate this with the family IV and with $k = 2$, this being perhaps the most interesting arithmetically. By choosing $\phi \in \mathcal{S}(\mathbb{R})$ so that $\phi(0) = 1$, $\phi(x) \geq 0$ and $\int_{-\infty}^{\infty} \phi(x) W(\mathcal{F}, x) dx$ is minimized (see [23]) we conclude from the density results in subsection 2.3 about family IV (which recall assume RH for automorphic L-functions) that for N, prime and large enough:

$$\frac{\#\{f \in H_2^+(N)|L(\frac{1}{2}, f) \neq 0\}}{\#\{f \in H_2^+(N)\}} > \frac{9}{16} \tag{4.3}$$

$$\frac{\#\{f \in H_2^-(N)|L'(\frac{1}{2}, f) \neq 0\}}{\#\{f \in H_2^-(N)\}} > \frac{15}{16} \tag{4.4}$$

and

$$\frac{|H_2(N)|}{2} + o(|H_2(N)|) \leq \sum_{f \in H_2(N)} ord(\frac{1}{2}, L(s, f)) < \frac{99}{100}|H_2(N)|, \tag{4.5}$$

where $ord(s_0, L(s, f))$ is the order of vanishing of $L(s, f)$ at $s = s_0$. Note that $|H_2(N)| \sim \frac{N}{12}$ and as Murty [33] shows (and this does not assume RH) that $|H_2^{\pm}(N)| \sim \frac{|H_2(N)|}{2}$, the lower bound in (4.5) is immediate. Concerning the upper bound in (4.5), Brumer [34] establishes such a result with 99/100 replaced by 3/2. One can reduce the 3/2 to 1 without appealing to the "off-diagonal" analysis of the last Section but to get anything below 1 already relies on extended ranges. A similar remark applies to (4.3), the off diagonal analysis allowing a lower bound bigger than 50%. This is significant as we will see below.

We can apply (4.3), (4.4), and (4.5) to the ranks of the Jacobians, $J_0(N)/\mathbb{Q}$, of the curves $X_0(N)$ (equal analytically to $\Gamma_0(N) \setminus \mathbb{H}$), by combining these results with known partial results to the Birch and Swinnerton-Dyer Conjecture (Kolyvagin [29] and Gross-Zagier [30]). Let $M_0(N)/\mathbb{Q}$ be the quotient of $J_0(N)$ considered by Merel [35]. It corresponds to the f's in $H_2^+(N)$ for which $L(\frac{1}{2}, f) \neq 0$ and is no doubt the largest quotient of $J_0(N)$ which is of rank zero. It is of great interest to know its size. Brumer [34] has computed these for $N \leq 10^4$ and based on his findings he conjectures that:

$$\lim_{N \to \infty} \frac{\dim M_0(N)}{|H_2^+(N)|} = 1 \tag{4.6}$$

$$\lim_{N \to \infty} \frac{\operatorname{rank} J_0(N)}{\dim J_0(N)} = \frac{1}{2}. \tag{4.7}$$

Note that the Density Conjectures for this family via (4.1) and (4.2), and [29] and [30] imply these Conjectures of Brumer. In the same way (4.3) and (4.4) imply (still under RH) that for N large:

$$\dim M_0(N) > \frac{9}{16} |H_2^+(N)| \tag{4.8}$$

and

$$\operatorname{rank} J_0(N) > \frac{15}{32} \dim(J_0(N)). \tag{4.9}$$

Moreover assuming the Birch and Swinnerton-Dyer Conjectures as well as (4.5) yields, that for N large

$$\frac{\dim J_0(N)}{2} + o(N) \leq \operatorname{rank} J_0(N) \leq \frac{99}{100} \dim J_0(N). \tag{4.10}$$

It is remarkable that the results (4.3), (4.4) and (4.5) can be established un-conditionally with almost as good quality. The techniques to achieve this are quite different and more sophisticated than those used for the density results for the families III, IV and V, though they both make use of the methods for averaging developed in [26]. In [36], Duke examines the averages of $L(\frac{1}{2}, f)$ and $L^2(\frac{1}{2}, f)$ over the family $H_2(N)$ and this allows him to show that at least $N/(\log N)^2$ of the $L(\frac{1}{2}, f)$'s are not zero. Introducing "mollifiers" and other tools into the analysis of averages of $L(\frac{1}{2}, f)$ and $L^2(\frac{1}{2}, f)$, Iwaniec and Sarnak [26] show the following:

$$\lim_{N \to \infty} \frac{\#\{f \in H_2^+(N)|L(\frac{1}{2}, f) \geq (\log N)^{-2}\}}{\#\{f \in H_2^+(N)\}} \geq \frac{1}{2}. \tag{4.11}$$

This unconditional result is rather close to the conditional result (4.3) and moreover the 50% is of fundamental significance. In [26] it is shown that if (4.11) holds with any $C > 1/2$ in place of $1/2$ on the right hand side, then there are no Siegel zeros! Of course the conditional result (4.3) is of no relevance here since tautologically the RH's imply that there are no Siegel zeros. Using variations of the techniques above among many other ideas

Kowalski and Michel [37] and independently VanderKam [38] have shown that (4.3) and (4.4) hold unconditionally for some positive constants on the right hand sides. All of these unconditional results when combined with [29] and [30] lead to corresponding unconditional results towards Brumer's Conjectures. In another work, Kowalski and Michel [39] established that the upper bound in (4.5) holds unconditionally with $99/100$ replaced by a large constant C.

5 Conclusion

P. Cohen once remarked to me that in a Colloquium talk, the first quarter should be understandable by everyone, the second by the experts, the third by the speaker and the end by no one. We now enter this final phase — at least as far as this speaker goes.

The results for function fields, the numerical experiments and the analytic results about densities all point convincingly to the fact that the distribution of zeros for families follow the $G(\mathcal{F})$ distributions. It is of course possible that $G(\mathcal{F})$ is simply an excellent model for predicting these densities. However based on what happens in the function field we believe that there is in fact a symmetry group in the global case which is the source of all of these phenomena. At a highly speculative level we expect (see [21]) that there is a natural spectral interpretation of the zeros of each $L(s, f)$ in terms of the eigenvalues of an operator $U(f)$ on a Hilbert space H (an interesting candidate for a spectral interpretation of the zeros of $L(s, \chi)$'s has been put forth by Connes [40]). Furthermore for one of our families $\mathcal{F}$ of such f's we expect that these $U(f)$'s can all be naturally defined on the same H. The symmetry $G(\mathcal{F})$ will then take the form that the corresponding operators $U(\mathcal{F})$ all preserve a corresponding structure on H (e.g. symplectic or orthogonal). The source of the distribution laws for families might then come from a grand "Chebotarev Theorem" asserting that as f varies over $\mathcal{F}$ with $c_f \leq X$, the $U(f)$'s become equidistributed in the corresponding space of operators. From this point of view it would follow from the Law of Large Numbers (2.5) of Lecture I and the universality of type II symmetric spaces, that for the typical member $f \in \mathcal{F}$, $L(s, f)$ satisfies the Montgomery-Odlyzko Law. That every $L(s, f)$ should satisfy this law, i.e. individually, is then special to the global L-functions (as mentioned before it does not apply in the function field or in the analogous Hamiltonian setting). In order to understand the symmetry of an individual $L(s, f)$ one should put the L-function in as small as possible family. For example, the Riemann Zeta function sits in the family I of Section 2.3 for which $G(\mathcal{F}) = Sp(\infty)$. We infer that in the proposed spectral interpretation of the zeros of the Riemann Zeta function the operator should preserve a natural symplectic structure!

The theme that there is a theory of families for global L-functions is a welcome one, since the proof by Deligne [19, 20] of the Weil Conjectures for zeta functions of varieties over finite fields (that is the generalization of the Riemann Hypothesis for function fields) uses the monodromy of families in a fundamental way.

To end we remark that one lesson that may be learned from this discussion on zeta functions that may apply to the case of Hamiltonians and in particular the Basic Conjectures is the following: In formulating the basic Conjecture for a family of Hamiltonians (i.e. that the measure theoretically typical member satisfies the Basic Conjecture) there should be a calculation which ensures that the family is large enough — just as the calculation of the monodromy being large, was crucial in the proof of (2.4).

6 Acknowledgments

I would like to thank my collaborators H. Iwaniec, N. Katz, W. Luo and Z. Rudnick, not only for the joint works which are mentioned above but also for their many ideas and insights which have been instrumental in shaping these ideas. Thanks also to M. Rubinstein whose numerical experimentation with zeros of L-functions are a major piece of the evidence for what we have described.

References

[1] B. Riemann, *Mon. der Berliner Akad.* 671 (1858/60).

[2] Titchmarsh, *The theory of the Riemann Zeta Function* (Calderon Press, Oxford, 1951).

[3] A. Odlyzko, The 10^{20} zero of the Riemann zeta function and its 70 million neighbours, preprint.

[4] E. Bogomolny and J. Keating, *Nonlinearity* **8**, 1115 (1995).

[5] H. Montgomery, *Proc. Symp. Pure Math.* **24**, 181 (1973).

[6] Z. Rudnick and P. Sarnak, *Duke Math. J.* **81**, 269 (1996).

[7] D. Goldston and H. Montgomery, in *Proc. Conference in Analytic Number Theory*, edited by A. Ghosh (Birkhauser, Boston, 1984), pp. 183–203.

[8] D. Hejhal, *IMRN* **1**, 293 (1994).

[9] F. Dyson, *J. Math. Phys.* **3**, 166 (1962).

[10] S. Gelbart and H. Jacquet, *Ann. Sci. Ecole Norm. Sup.* 4^e **Serie II**, 471 (1978).

[11] H. Jacquet, *Proc. Sym. Pure Math.* **2**, 63 (1979).

[12] J. P. Serre, *A course in arithmetic* (Springer, N.Y., 1972).

[13] R. Rumely, *Math. of Comp.* **61**, 415 (1993).

[14] M. Rubinstein, in preparation.

[15] E. Artin, *Math. Zeit.* **19**, 153 (1924).

[16] C. Schmit, *Chaos and Quantum Physics (in Les Houches, conference)* (North Holland, Amsterdam, 1991), pp. 333–369.

[17] A. Weil, *Proc. Nat. Acad. Sci., USA* **27**, 345 (1941).

[18] N. Katz and P. Sarnak, *Random matrices, Frobenius eigenvalues and monodromy* (AMS, Rhode Island, 1998), to appear.

[19] P. Deligne, *Publ. IHES* **48**, 273 (1974).

[20] P. Deligne, *Publ. IHES* **52**, 313 (1981).

[21] N. Katz and P. Sarnak, Zeros of zeta functions, their spacings and their spectral nature, 1997, preprint.

[22] A. Ozluk and C. Snyder, *Bull. Aust. Math. Soc.* **47**, 307 (1993).

[23] H. Iwaniec, W. Luo, and P. Sarnak, Low Lying zeros for families of L-functions, in preparation.

[24] Atkin and J. Lehner, *Math. Ann.* **185**, 134 (1970).

[25] G. Shimura, *Proc. London Math. Soc.* **31**, 79 (1975).

[26] H. Iwaniec and P. Sarnak, The non-vanishing of central values of automorphic L-functions and Siegel's zeros, 1997, the nonvanishing of central values of automorphic L-functions and Siegel's zeros, preprint 1997.

[27] M. Berry, *Proc. Royal. Soc. London* **A 400**, 229 (1985).

[28] W. Duke, J. Friedlander, and H. Iwaniec, *Sieve Methods, Exponential Sums and their Applications in Number Theory* (Cambridge Univ. Press, N.Y., 1997), pp. 109–115.

[29] V. A. Kolyvagin and D. Lugachev, *Leningrad Math J.* **1**, 1229 (1990).

[30] B. Gross and D. Zagier, *Invent. Math.* **84**, 225 (1986).

[31] J. Waldspurger, *J. Math. Pures Appl.* **60**, 375 (1981).

[32] R. Phillips and P. Sarnak, *Invent. Math.* **80**, 339 (1985).

[33] R. Murty, *CMS Conf. Proc.* **15**, 263 (1995).

[34] A. Brumer, *Asterisque* **228**, 41 (1995).

[35] L. Merel, *Invent. Math.* **124**, 437 (1996).

[36] W. Duke, *Invent. Math.* **119**, 165 (1995).

[37] E. Kowalski and P. Michel, Sur les zeros des fonctions L de formes automorphes, preprint.

[38] J. Vanderkam, The rank of $J_0(N)$, 1997, preprint.

[39] E. Kowalski and P. Michel, Sur le rang de $J_0(q)$, 1997, preprint.

[40] A. Connes, *C. R. Acad. S. C. Paris* **2**, 1231 (1996).

Character formulas for tilting modules over quantum groups at roots of one

Wolfgang Soergel

Universität Freiburg
Mathematisches Institut
Eckerstraße 1
D-79104 Freiburg
Germany
soergel@mathematik.uni-freiburg.de

1 Motivation

Let k be an algebraically closed field, $n \geq 1$ an integer. Our motivating problem is the study of the irreducible representations of the symmetric group $\mathcal{S}_n$ over k, in other words of the irreducible modules over the group ring $k\mathcal{S}_n$. In case $\operatorname{char} k = 0$ their classification is well known, we have a bijection

$$\operatorname{Irr} k\mathcal{S}_n \leftrightarrow \{\text{Partitions of } n\}.$$

In case $\operatorname{char} k = p > 0$ the classification is less well known, we have a bijection

$$\operatorname{Irr} k\mathcal{S}_n \leftrightarrow \left\{ \begin{array}{l} \text{Partitions of } n \text{ which do not} \\ \text{have } p \text{ or more equal pieces} \end{array} \right\}.$$

The next question is, given such a partition, to determine the dimension of the corresponding irreducible representation of the symmetric group. In case $\mathrm{char}k = 0$, these dimensions are again well known. We just have to display our partition as a Young diagram and count the standard tableaux of this shape, i.e. the dimension in question is the number of possibilities we have to enumerate the boxes of our Young diagram such that each row and column is increasing. For all this see [10]. In case $\mathrm{char}k = p > 0$, the dimensions are not even known for all partitions of n into three pieces.

This is the problem where tilting modules suggest a new line of attack [6, 8, 9, 17]. Namely consider a finite dimensional vector space V over k of dimension $\dim_k V \geq n$. Then "Schur-Weyl duality" (see [4] or 2.9 below for the case of arbitrary characteristic) says that

$$\mathrm{End}_{\mathrm{GL}(V)} V^{\otimes n} = k \mathcal{S}_n$$

with the symmetric group permuting the factors of a tensor in $V^{\otimes n}$.

If $\mathrm{char}k = 0$, we know that $V^{\otimes n}$ decomposes into a direct sum of irreducible subrepresentations under $\mathrm{GL}(V)$. In general, this is no longer true. But suppose

$$V^{\otimes n} \cong \bigoplus_{\lambda \in \Lambda} T(\lambda)^{m(\lambda)}$$

is a decomposition of the $\mathrm{GL}(V)$-module $V^{\otimes n}$ into indecomposable (not in general irreducible) direct summands, where the isomorphism classes of indecomposables $T(\lambda)$ appearing are parametrized by a suitable set Λ and occur with multiplicity $m(\lambda) > 0$. Then it is clear from abstract algebra that the quotient of $k\mathcal{S}_n$ by its Jacobson radical is just a product of matrix algebras of the form

$$k\mathcal{S}_n / \mathrm{rad}k\mathcal{S}_n \cong \Pi_{\lambda \in \Lambda} M(m(\lambda) \times m(\lambda), k)$$

and we get a bijection

$$\mathrm{Irr}k\mathcal{S}_n \qquad \longleftrightarrow \qquad \Lambda$$
$$\begin{pmatrix} \text{some irreducible} \\ \text{of dimension } m(\lambda) \end{pmatrix} \qquad \mapsto \qquad \lambda.$$

Therefore it is of the utmost importance to determine the decomposition of $V^{\otimes n}$ into indecomposables. In the next section we will following [6] define what is a "tilting module" for $G = \mathrm{GL}(V)$ or more generally for an arbitrary reductive algebraic group G. We will see that the indecomposable tilting modules are classified by their highest weights, and that all the summands $T(\lambda)$ of $V^{\otimes n}$ above are actually tilting modules.

Since the character of $V^{\otimes n}$ is known, we just need to determine the character of the indecomposable tilting module for any given highest weight to obtain the looked-for decomposition of $V^{\otimes n}$. We can only solve the analogous problem with "algebraic groups in positive characteristic" replaced by "quantum groups at roots of one", and conjecture (see [1]) that if the highest weight is not too big, the quantized problem has the same solution as the original one. In the hope that the reader is now sufficiently motivated let me go on to define tilting modules.

2 Tilting modules for reductive algebraic groups

Let k be an algebraically closed field, $G \supset B \supset H$ a connected reductive algebraic group over k, a Borel subgroup and a maximal torus (see [11] for all this foundational material). The inclusion $H \subset B$ admits a unique left inverse $B \twoheadrightarrow H$ which allows us to consider every character of H as a character of B. For λ in the character lattice $X = X(H)$ of H we consider the induced representation

$$
\begin{aligned}
\nabla(\lambda) &= \operatorname{ind}_B^G k_\lambda \\
&= \left\{ f : G \to k \, \middle| \, \begin{array}{l} f \text{ is algebraic and satisfies} \\ f(xb) = \lambda(b)^{-1} f(x) \quad \forall x \in G, b \in B \end{array} \right\}
\end{aligned}
$$

with the action of $g \in G$ on $f \in \nabla(\lambda)$ defined by $(gf)(x) = f(g^{-1}x)$. If we let $R \subset X$ be the root system of G, take as positive roots the complement $R^+ = R - R(B)$ of the roots of B, and let $X^+ \subset X$ be the corresponding dominant weights, then we have $\nabla(\lambda) \neq 0$ iff $\lambda \in X^+$.

In case char$k = 0$ our $\nabla(\lambda)$ is precisely the simple representation of G with highest weight λ. In general, the character of $\nabla(\lambda)$ is still given by the Weyl character formula, but the $\nabla(\lambda)$ are no longer simple for all λ, as we can readily see from the example $G = \mathrm{GL}(2,k)$: In this case letting ρ be half the positive root we can identify the representation $\nabla(n\rho)$ with the obvious representation $k[X,Y]^n$ of $\mathrm{GL}(2,k)$ on polynomials in two variables homogeneous of degree n, and for char$k = p$ obviously $kX^p + kY^p \subset k[X,Y]^p$ is an invariant subspace. We can further see that this subspace has no invariant complement, so $\nabla(p\rho)$ is not completely reducible if char$k = p > 0$.

Definition 2.1. [6] A rational representation T of G is called a "tilting module" if and only if both T and its dual T^* admit a finite ∇-flag, i.e. a finite filtration by G-stable subspaces such that all successive subquotients are isomorphic to some $\nabla(\lambda)$. In particular a tilting module by definition is always of finite dimension over k.

Remark 2.2. In the terminology of Donkin (due to Ringel [18]) such representations would in fact be called a "partial tilting modules". However common usage is now to just call them tilting modules.

Examples 2.3. The trivial representation $k = \nabla(0)$ is always tilting. If more generally $\nabla(\lambda)$ is simple, one can show that $\nabla(\lambda)^* \cong \nabla(-w_\circ\lambda)$ for $w_\circ$ the longest element of the Weyl group, hence every simple $\nabla(\lambda)$ is tilting. As a special case of this we see that V is a tilting module for $GL(V)$.

To have a non-trivial example let us consider the tensor product $\nabla(\rho) \otimes \nabla(p\rho - \rho)$ for $G = GL(2, k)$, char$k = p$. This is certainly selfdual and one may see by hand that it fits into a short exact sequence

$$\nabla(p\rho - 2\rho) \hookrightarrow \nabla(\rho) \otimes \nabla(p\rho - \rho) \twoheadrightarrow \nabla(p\rho).$$

Therefore $\nabla(\rho) \otimes \nabla(p\rho - \rho)$ is a tilting module for $G = GL(2, k)$.

The following theorem collects some of the reasons, why tilting modules are interesting.

Theorem 2.4.   *1. Every summand of a tilting module is tilting.*

   2. The tensor product of two tilting modules is tilting.

   3. The indecomposable tilting modules are classified by their highest weights, more precisely we have a bijection

$$\left\{ \begin{array}{c} \textit{indecomposable tilting modules,} \\ \textit{up to isomorphism} \end{array} \right\} \quad \leftrightarrow \qquad X^+$$
$$\qquad\qquad T \qquad\qquad\qquad \mapsto \quad \textit{"the highest weight of } T\textit{"}$$

Notation 2.5. We denote by $T(\lambda)$ the indecomposable tilting module with highest weight λ.

Proof. Let us give some indications on where this theorem comes from. Part (2) follows easily from the following theorem due in full generality to [16]:

Theorem 2.6. *For all $\lambda, \mu \in X^+$ the tensor product $\nabla(\lambda) \otimes \nabla(\mu)$ admits a ∇-flag.*

To explain part (1), let us define $\Delta(\lambda) = \nabla(-w_\circ\lambda)^*$ for every $\lambda \in X$. (The parametrization of the $\Delta(\lambda)$ is just set up in such a way that $\Delta(\lambda)$ and $\nabla(\lambda)$ have the same character.) Let Ext_G denote extensions in the category of all rational representations of G. Part (1) then follows easily from the following result of Donkin [5], (see also [Ja], II, 4.16).

Proposition 2.7. *For a finite dimensional rational representation T of G the following are equivalent:*

   1. T admits a ∇-flag.

   2. $\mathrm{Ext}_G^1(\Delta(\lambda), T) = 0 \quad \forall \lambda \in X^+$.

Part (3) of the theorem can be proved in a very general context, as is explained in the next section following [18]. For this we use the usual partial order on X^+ given by $\lambda \geq \mu$ iff $\lambda \in \mu + \mathbb{N}R^+$ and then only need to know that $\text{Hom}_G(\Delta(\lambda), \Delta(\mu)) = 0$ unless $\lambda \leq \mu$, explicitely verifying explicitly$\text{Ext}^1_G(\Delta(\lambda), \Delta(\mu))$ vanishes unless $\lambda < \mu$, and both are finite dimensional for all λ, μ. This can be deduced from results in [Ja] without too much difficulty. □

Remark 2.8. From the theorem and the fact that V is tilting for $\text{GL}(V)$ it is clear that the Krull-Schmid-decomposition of the $\text{GL}(V)$-module $V^{\otimes n}$ has to be of the form

$$V^{\otimes n} = \bigoplus_{\lambda \in X^+} T(\lambda)^{m(\lambda)}$$

where the $m(\lambda)$ are suitable multiplicities $m(\lambda) \geq 0$ and the $T(\lambda)$ are our indecomposable tilting modules from 2.5. Already for $G = \text{GL}(3, k)$ however the characters of the $T(\lambda)$ are not all known, and what's worse, there isn't even a general conjecture. If λ is in the "lowest p^2-alcove", then $T(\lambda)$ is conjectured to have the same character as its quantum analogue, see [1].

Remark 2.9. Let me explain how the above results imply the fundamental formula $\text{End}_{\text{GL}(V)} V^{\otimes n} = k\mathcal{S}_n$ (if k is infinite and $\dim_k V \geq n$). that $\text{End}_{k\mathcal{S}_n} V^{\otimes n}$ is the k-linear subspace of $\text{End}_k V^{\otimes n}$ generated by the image of $\text{GL}(V)$. If $\text{char} k = 0$, then $V^{\otimes n}$ is a semisimple $k\mathcal{S}_n$-module and thus by the Jacobson density theorem $k\mathcal{S}_n$ surjects onto the bicommutant of its image in $\text{End}_k V^{\otimes n}$, i.e. we have a surjection $k\mathcal{S}_n \twoheadrightarrow \text{End}_{\text{GL}(V)} V^{\otimes n}$.

For $\text{char} k$ is arbitrary it is still clear that $k\mathcal{S}_n \to \text{End}_{\text{GL}(V)} V^{\otimes n}$ is an injection in case $\dim_k V \geq n$. The problem is surjectivity. For this we will count dimensions and just have to show that the dimension of $\text{End}_{\text{GL}(V)} V^{\otimes n}$ is independent of the ground field k. For this in turn we need to know that $\dim_k \text{Hom}_G(\Delta(\lambda), \nabla(\mu)) = \delta_{\lambda, \mu}$ and $\text{Ext}^1_G(\Delta(\lambda), \nabla(\mu)) = 0$ for all λ, $\mu \in X^+$. (The first statement can be found in say [11], the second one also is a special case of 2.7 above.)

Then if T has a finite Δ-flag where $\Delta(\lambda)$ occurs with multiplicity $(T : \Delta(\lambda))$ and T' has a finite ∇-flag where $\nabla(\lambda)$ occurs with multiplicity $(T' : \nabla(\lambda))$, we get

$$\dim_k \text{Hom}_G(T, T') = \sum_\lambda (T : \Delta(\lambda))(T' : \nabla(\lambda)).$$

In particular, we get

$$\dim_k \text{End}_{\text{GL}(V)} V^{\otimes n} = \sum_\lambda (V^{\otimes n} : \Delta(\lambda))(V^{\otimes n} : \nabla(\lambda))$$

which is indeed independent from the ground field k. As an aside, we see from comparing characters that $(T : \Delta(\lambda)) = (T : \nabla(\lambda))$ if T is tilting.

3 Existence and unicity of tilting modules in an abstract context

Suppose k is a field, $\mathcal{C}$ an Abelian k-category, and $\{\Delta(\lambda)\}_{\lambda \in \Lambda}$ a collection of indecomposable objects of $\mathcal{C}$ parametrized by a partially ordered set $(\Lambda, \leq)$, such that the following conditions are satisfied:

1. $\mathrm{Hom}(\Delta(\lambda), \Delta(\mu)) = 0$ unless $\lambda \leq \mu$.

2. $\mathrm{Ext}^1_{\mathcal{C}}(\Delta(\lambda), \Delta(\mu)) = 0$ unless $\lambda < \mu$.

3. $\dim \mathrm{Hom}_{\mathcal{C}}(\Delta(\lambda), \Delta(\mu)) < \infty$, $\dim \mathrm{Ext}^1_{\mathcal{C}}(\Delta(\lambda), \Delta(\mu)) < \infty$ $\forall \lambda, \mu$.

4. For $\lambda \in \Lambda$ there are only finitely many elements below it.

In this situation we have

Proposition 3.1. *[18] For every $\lambda \in \Lambda$ there exists a unique (up to non-unique isomorphism) indecomposable object $T = T(\lambda) \in \mathcal{C}$, which satisfies the following two conditions:*

(a) $\mathrm{Ext}^1_{\mathcal{C}}(\Delta(\nu), T) = 0$ $\forall \nu \in \Lambda$.

(b) T admits a finite Δ-flag starting with $\Delta(\lambda) \subset T$.

Remark 3.2. By a finite Δ-flag of an object $T \in \mathcal{C}$ we mean a filtration $0 = T_0 \subset T_1 \subset T_2 \subset \ldots T_r = T$ such that $T_i/T_{i-1} \cong \Delta(\lambda_i)$ for suitable $\lambda_i \in \Lambda$.

Definition 3.3. $T(\lambda)$ might be called the indecomposable tilting module with parameter λ, but in this generality the terminology is not commonly used.

Proof. Existence: We prove the existence of $T(\lambda)$ by induction on the number of elements below λ. If λ is already minimal, we can take $T(\lambda) = \Delta(\lambda)$. If not, we find by condition (4) a minimal $\mu \in \Lambda$ with $\mu < \lambda$. By induction, we know there exists $\tilde{T} \in \mathcal{C}$ such that $\mathrm{Ext}^1_{\mathcal{C}}(\Delta(\nu), \tilde{T}) = 0$ for all $\nu \neq \mu$ and that $\tilde{T}$ admits a finite Δ-flag with subquotients $\Delta(\nu)$, $\nu \neq \mu$. If $\mathrm{Ext}^1_{\mathcal{C}}(\Delta(\mu), \tilde{T}) = 0$ we can take $\tilde{T} = T(\lambda)$ and are through. If not, we have at least $\dim_k \mathrm{Ext}^1_{\mathcal{C}}(\Delta(\mu), \tilde{T}) < \infty$ by condition (3). Put $\tilde{T} = T_0$, choose a nonzero element e of this Ext-group, and represent it by a short exact sequence

$$T_0 \hookrightarrow T_1 \twoheadrightarrow \Delta(\mu).$$

A segment of the corresponding long exact sequence of Ext-groups is

$$\mathrm{Hom}_{\mathcal{C}}(\Delta(\mu), \Delta(\mu)) \to \mathrm{Ext}^1_{\mathcal{C}}(\Delta(\mu), T_0) \to \mathrm{Ext}^1_{\mathcal{C}}(\Delta(\mu), T_1) \to 0,$$

the last zero since we have $\text{Ext}^1_{\mathcal{C}}(\Delta(\mu), \Delta(\mu)) = 0$ by condition (1). Now the boundary map maps id $\in \text{Hom}_{\mathcal{C}}(\Delta(\mu), \Delta(\mu))$ to our nonzero element e. Thus $\dim_k \text{Ext}^1_{\mathcal{C}}(\Delta(\mu), T_1) < \dim_k \text{Ext}^1_{\mathcal{C}}(\Delta(\mu), T_0)$ and by other long exact sequences still $\text{Ext}^1_{\mathcal{C}}(\Delta(\nu), T_1) = 0$ if $\nu \neq \mu$. Continuing this way, we construct inductively $T_2, T_3, \dots$ until we arrive at some T_r that fits into a short exact sequence

$$\tilde{T} \hookrightarrow T_r \twoheadrightarrow \Delta(\mu)^r$$

and satisfies $\text{Ext}^1_{\mathcal{C}}(\Delta(\nu), T_r) = 0 \quad \forall \nu \in \Lambda$. In case $\text{End}_{\mathcal{C}} \Delta(\mu) = k$ it can be shown that T_r is actually indecomposable, but in our general situation this need not be true. To circumvent this problem we rather show that if m is the smallest possible integer such that there exists a short exact sequence

$$\tilde{T} \hookrightarrow T \twoheadrightarrow \Delta(\mu)^m$$

with T satisfying $\text{Ext}^1_{\mathcal{C}}(\Delta(\nu), T) = 0 \quad \forall \nu$, then this T is indecomposable and hence our looked-for $T(\lambda)$. Indeed, we can describe $\tilde{T}$ as the biggest subobject of T killed by all morphisms $T \to \Delta(\mu)$, and hence any decomposition $T = T' \oplus T''$ induces a decomposition $\tilde{T} = \tilde{T}' \oplus \tilde{T}''$. Since $\tilde{T}$ is indecomposable, we may assume $\tilde{T} = \tilde{T}'$, $\tilde{T}'' = 0$ and get $T'/\tilde{T} \oplus T'' \cong \Delta(\mu)^m$. Now the Krull-Schmid theorem implies $T'/\tilde{T} \cong \Delta(\mu)^n$ with $n \leq m$, but by minimality of m we have necessarily $n = m$, hence $T'' = 0$.

Unicity: Since T is indecomposable, $\text{End}_{\mathcal{C}} T$ admits no idempotents except zero and one. Since T has a finite Δ-flag, $\text{End}_{\mathcal{C}} T$ is of finite dimension over k by (2). Together this implies that every element of $\text{End}_{\mathcal{C}} T$ is either nilpotent or an isomorphism, by the Lemma of Fitting.

Suppose now $T, T' \in \mathcal{C}$ are two indecomposables satisfying (a) and (b). Consider the diagram

$$
\begin{array}{ccccc}
\Delta(\lambda) & \hookrightarrow & T & \twoheadrightarrow & \text{coker} \\
\| & & & & \\
\Delta(\lambda) & \hookrightarrow & T' & \twoheadrightarrow & \text{coker}'
\end{array}
$$

with short exact horizontals. By (b) coker has a Δ-flag, thus by (a) we have $\text{Ext}^1_{\mathcal{C}}(\text{coker}, T') = 0$, thus the restriction $\text{Hom}_{\mathcal{C}}(T, T') \to \text{Hom}_{\mathcal{C}}(\Delta(\lambda), T')$ is a surjection and we can find $\alpha : T \to T'$ inducing the identity on $\Delta(\lambda)$. Similarly we also find $\beta : T' \to T$ inducing the identity on $\Delta(\lambda)$. Since $\beta \circ \alpha \in \text{End}_{\mathcal{C}} T$ is not nilpotent, it has to be an isomorphism, and the same holds for $\alpha \circ \beta$. We conclude $T \cong T'$. $\qquad\square$

The multiplicity of $\Delta(\nu)$ as subquotient in a Δ-flag of $T(\lambda)$ will be denoted by $(T(\lambda) : \Delta(\nu))$. This number is independent of the Δ-flag. Indeed, if we enumerate the elements below λ as $\lambda_1, \dots, \lambda_n$ such that $\lambda_i > \lambda_j \Rightarrow i < j$, in particular $\lambda_i = \lambda$ and λ_n is minimal, then by the vanishing of the relevant

Ext^1_C any Δ-flag can be transformed without changing its multiplicities to a Δ-flag where first come the subquotients $\Delta(\lambda_1)$, then the $\Delta(\lambda_2)$ etc. It is then clear from our inductive construction of $T(\lambda)$ that all multiplicities are well determined.

4 Character formulas for indecomposable tilting modules over graded Lie algebras

We will explain next how to determine the characters of indecomposable tilting modules for quantum groups at roots of unity or equivalently [12, 13] how to solve the translated problem in a suitable category $\mathcal{O}$ of representations of an affine Lie algebra. First we will work with an arbitrary $\mathbb{Z}$-graded Lie-algebra $\mathfrak{g} = \bigoplus_{i \in \mathbb{Z}} \mathfrak{g}_i$ over the field k subject only to the following three conditions

1. $\dim_k \mathfrak{g}_i < \infty \quad \forall i \in \mathbb{Z}$.

2. $\mathfrak{g}$ is generated by $\mathfrak{g}_{-1}, \mathfrak{g}_0, \mathfrak{g}_1$.

3. There exists a character $\gamma : \mathfrak{g}_0 \to k$ such that $\operatorname{tr}(\operatorname{ad}X \operatorname{ad}Y : \mathfrak{g}_0 \to \mathfrak{g}_0) = \gamma([X, Y])$ for all $X \in \mathfrak{g}_1$, $Y \in \mathfrak{g}_{-1}$.

Let us put $\mathfrak{g}_{<0} = \mathfrak{n}$, $\mathfrak{g}_{\geq 0} = \mathfrak{b}$ and denote the enveloping algebras of $\mathfrak{g}, \mathfrak{n}, \mathfrak{b}$ by U, N, B. Let $\mathcal{M}$ denote the category of all $\mathbb{Z}$-graded $\mathfrak{g}$-modules, which are graded free of finite rank over N. The following result is due to Archipov [3] and can also be found in [19].

Theorem 4.1. *There exists an equivalence of categories $\mathcal{M} \to \mathcal{M}^{\mathrm{opp}}$ such that short exact sequences of $\mathfrak{g}$-modules on both sides correspond, and that $U \otimes_B E$ gets mapped to $U \otimes_B (k_{-\gamma} \otimes E^*)$ for every finite dimensional $\mathbb{Z}$-graded representation E of $\mathfrak{g}_0$.*

Proof. I will only give some indications on how this result may be proven, more details can be found in [19]. Denote for a $\mathbb{Z}$-graded space $M = \oplus M_i$ by $M^{\circledast}$ its graded dual, so $(M^{\circledast})_i = (M_{-i})^*$. We make $N^{\circledast}$ into a $\mathbb{Z}$-graded N-bimodule in the most obvious way, the left action of N on itself giving rise to the right action on $N^{\circledast}$ and vice versa. The key ingredient to our equivalence will be a very peculiar $\mathbb{Z}$-graded U-bimodule S, the so-called "semi-regular bimodule", whose existence is assured by the following

Proposition 4.2. *There exists a $\mathbb{Z}$-graded U-bimodule $S = S_\gamma$ along with an inclusion $\iota : N^{\circledast} \hookrightarrow S$ of $\mathbb{Z}$-graded N-bimodules such that the following holds:*

1. *The map $U \otimes_N N^{\circledast} \to S$, $u \otimes f \mapsto u\iota(f)$ is a bijection.*

2. *The map* $N^{\circledast} \otimes_N U \to S$, $f \otimes u \mapsto \iota(f)u$ *is a bijection.*

3. *Up to a twist by γ the inclusion $\iota : N^{\circledast} \hookrightarrow S$ is compatible with the adjoint action of $\mathfrak{g}_0$ on both spaces, more precisely $\iota(f \circ \mathrm{ad}H) + (\mathrm{ad}H)\iota(f) = \iota(f)\gamma(H)$ for all $H \in \mathfrak{g}_0$ and $f \in N^{\circledast}$.*

This Proposition can be checked by brute force, but I still don't know a good proof. Anyhow, we can now write down the equivalence of categories claimed by the theorem as the functor $M \mapsto (S \otimes_U M)^{\circledast}$, and it is not difficult to check that this has the requested properties. □

Suppose now we are working over a ground field k of characteristic zero and $\mathfrak{g}$ is semisimple under the adjoint action of $\mathfrak{g}_0$. Let $\mathcal{O}$ be the category of all $\mathbb{Z}$-graded $\mathfrak{g}$-modules which are locally finite over $\mathfrak{g}_{\geq 0}$ and semisimple as modules over $\mathfrak{g}_0$. Let Λ be the set of isomorphism classes of finite dimensional simple $\mathbb{Z}$-graded $\mathfrak{g}_0$-modules. Any $E \in \Lambda$ is necessarily concentrated in just one degree. For $E \in \Lambda$ the Verma module $\Delta(E) = U \otimes_B E$ admits a unique simple quotient $L(E)$, and in this way Λ parametizes the simple objects of $\mathcal{O}$ up to isomorphism. Adapting the arguments of section 3 to our situation, we can prove

Theorem 4.3. *For every $E \in \Lambda$ there exists a unique (up to non-unique isomorphism) indecomposable object $T = T(E) \in \mathcal{O}$ which satisfies the following two conditions:*

1. $\mathrm{Ext}^1_{\mathcal{O}}(\Delta(F), T(E)) = 0 \quad \forall F \in \Lambda.$

2. *T admits a Δ-flag starting with $\Delta(E)$, i.e. a filtration $0 = T_0 \subset T_1 \subset T_2 \subset \ldots$ such that $T_1 \cong \Delta(E)$, $\bigcup T_i = T$ and $T_i/T_{i-1} \cong \Delta(F_i)$ for suitable $F_i \in \Lambda$.*

Certainly we call $T(E)$ the tilting module with parameter E. The main theorem is a formula for the multiplicity $(T(E) : \Delta(F))$ of $\Delta(F)$ as subquotient in a Δ-flag of $T(E)$ as above. To formulate the theorem in this generality, we have to introduce the module

$$\nabla(E) = (U \otimes_{U(\mathfrak{g}_{\leq 0})} E^*)^{\circledast}$$

which is the correct generalization of what is known as the dual Verma in the Kac-Moody set-up. Let $[\nabla(E) : L(F)]$ denote the Jordan-Hölder-multiplicity of $L(F)$ in $\nabla(E)$. The "abstract character formula for tilting modules" then says

Theorem 4.4. $(T(E) : \Delta(F)) = [\nabla(k_{-\gamma} \otimes F^*) : L(k_{-\gamma} \otimes E^*)]$ *for all $E, F \in \Lambda$.*

Proof. We will only prove this under the additional assumption that the simple object $L = L(k_{-\gamma} \otimes E^*)$ admits an indecomposable projective cover $P = P(k_{-\gamma} \otimes E^*)$ in $\mathcal{O}$. (The general case is more or less the same, details can be found in [19].) Under our assumption one proves as usual that P admits a finite Δ-flag ending with $\Delta(k_{-\gamma} \otimes E^*)$, and a suitably general form of the reciprocity principle tells us that the multiplicities in this Δ-flag are given by $(P : \Delta(F)) = [\nabla(F) : L]$ for $F \in \Lambda$.

Now we take a second look at our functor $\mathcal{M} \overset{\sim}{\to} \mathcal{M}^{\mathrm{opp}}$ to realize that it also gives an equivalence $\mathcal{M} \cap \mathcal{O} \overset{\sim}{\to} (\mathcal{M} \cap \mathcal{O})^{\mathrm{opp}}$. This in turn means that our functor transforms P into an indecomposable object T that has a Δ-flag starting with $\Delta(E)$ and satisfies $\mathrm{Ext}^1_{\mathcal{O}}(\Delta(F), T) = 0 \quad \forall F \in \Lambda$. In other words, $P = P(k_{-\gamma} \otimes E^*)$ gets transformed into our tilting module $T = T(E)$, and we can calculate

$$
\begin{aligned}
(T(E) : \Delta(F)) &= (P(k_{-\gamma} \otimes E^*) : \Delta(k_{-\gamma} \otimes F^*)) \\
&= [\nabla(k_{-\gamma} \otimes F^*) : L(k_{-\gamma} \otimes E^*)].
\end{aligned}
$$

$\square$

5   The case of quantum groups

If we run the abstract formula from Theorem 4.4 for the special case of affine Lie algebras, we see that the character formulas for tilting modules in negative level are determined by the character formulas for simple modules in positive level and vice versa. Now results of Kazhdan and Lusztig [12,13] tell us how to relate representations of quantum groups at roots of one to category $\mathcal{O}$ at negative level for affine Lie algebras. Putting all this together, we see that the character formulas of [14] for simple highest weight modules in positive level lead to character formulas for tilting modules for quantum groups at roots of one.

This however only works fine in simply laced cases. In general we would need the extension of the results of [14] to non-integral highest weights, which is still missing from the literature. The (partially conjectural) outcome is a formula expressing the characters of tilting modules in terms of Kazhdan-Lusztig polynomials, see [19] for details and [20] for motivation.

In the special case that our quantum group U is "quantized $\mathrm{GL}(V)$", we have by [7] 3.6 the quantized Schur-Weyl duality

$$
\mathcal{H} = \mathrm{End}_U V^{\otimes n}
$$

where $\mathcal{H}$ denotes the "quantization" of $k\mathcal{S}_n$, i.e. the Hecke algebra of the symmetric group. The same arguments as in section 1 now tell us in which way the character formulas for indecomposable tilting modules over U determine the dimensions of all simple modules over $\mathcal{H}$. However I do not yet see how to obtain from there the formulas for these dimensions conjectured in [15] and proved by Araki [2] and Grojnowski.

References

[1] Henning Haahr Andersen. Tilting modules for algebraic groups. to appear in Proceedings of NATO ASI at Newton Institute, 1997.

[2] S. Araki. On the decomposition numbers of the hecke algebra $g(m, 1, n)$. appeared, but where?, 1996.

[3] Sergej M. Arkhipov. Semiinfinite cohomology of associative algebras and bar duality. Preprint q-alg/9602013, 1996.

[4] Roger W. Carter and George Lusztig. On the modular representations of the general linear and symmetric groups. *Mathematische Zeitschrift*, 136:193–242, 1974.

[5] Stephen Donkin. A filtration for rational modules. *Mathematische Zeitschrift*, 177:1–8, 1981.

[6] Stephen Donkin. On tilting modules for algebraic groups. *Mathematische Zeitschrift*, 212:39–60, 1993.

[7] Jie Du. A note on quantized weyl reciprocity at roots of unity. *Algebra Colloq.*, 2(4):363–372, 1995.

[8] Karin Erdmann. Symmetric groups and quasi-hereditary algebras. In V. Dlab and L. L. Scott, editors, *Proc. Conf. Finite Dimensional Algebras and Related Topics*, pages 123–161. Kluwer, 1994.

[9] Karin Erdmann. Tensor products and dimensions of simple modules for symmetric groups. *manuscripta mathematica*, 88:357–386, 1995.

[10] G. D. James. *The Representation Theory of the Symmetric Groups*, volume 682 of *Lecture Notes in Mathematics*. Springer, 1978.

[11] Jens C. Jantzen. *Representations of Algebraic Groups*, volume 131 of *Pure and applied mathematics*. Academic Press, 1987.

[12] David Kazhdan and George Lusztig. Tensor structures arising from affine Lie algebras, I, II. *Journal of the AMS*, 6:905–1011, 1993.

[13] David Kazhdan and George Lusztig. Tensor structures arising from affine Lie algebras, III, IV. *Journal of the AMS*, 7:335–453, 1994.

[14] Masaki Kashiwara and Toshiyuki Tanisaki. Kazhdan-Lusztig conjecture for affine Lie algebras with negative level. *Duke Mathematical Journal*, 77:21–62, 1995.

[15] Alain Lascoux, Bernard Leclerc, and Jean-Yves Thibon. Hecke algebras at roots of unity and crystal bases of quantum affine algebras. *Commun. Math. Phys.*, 181(1):205–263, 1996.

[16] Olivier Mathieu. Filtrations of g-modules. *Ann. Scient. Éc. Norm. Sup.*, 23:625–644, 1990.

[17] Olivier Mathieu. On the dimension of some modular irreducible representations of the symmetric group. *Lett. Math. Phys.*, 38(1):23–32, 1996.

[18] Claus Michael Ringel. The category of modules with good filtrations over a quasi-hereditary algebra has almost split sequences. *Mathematische Zeitschrift*, 208:209–223, 1991.

[19] Wolfgang Soergel. Charakterformeln für Kipp-Moduln über Kac-Moody-Algebren. *Representation Theory (An electronic Journal of the AMS)*, 1:115–132, 1997.

[20] Wolfgang Soergel. Kazhdan-Lusztig-Polynome und eine Kombinatorik für Kipp-Moduln. *Representation Theory (An electronic Journal of the AMS)*, 1:37–68, 1997.

VOEVODSKY'S PROOF OF THE MILNOR CONJECTURE

ANDREI SUSLIN

Introduction

Let F be a field. The Milnor's ring of F is defined as a factor-ring of the tensor algebra $T(F^*)$ of the multiplicative group of F modulo a homogenous ideal generated by tensors of the form $a \otimes (1-a)$ (with $a \in F^* \setminus 1$). Thus

$$K_*^M(F) = K_0^M(F) \oplus K_1^M(F) \oplus K_2^M(F) \oplus \dots$$

is a graded ring, whose n-th homogenous component $K_n^M(F)$ coincides with an abelian group generated by symbols $\{a_1, ..., a_n\}$ $(a_i \in F^*)$ which are subject to two relations

(1) Multiplicativity in each variable
(2) $\{a_1, ..., a_n\} = 0$ provided that $a_i + a_{i+1} = 1$ for some i.

Obviously $K_0^M(F) = \mathbb{Z}$, $K_1^M(F) = F^*$, furthermore the group $K_2^M(F)$ coincides with Quillen's $K_2(F)$ in view of the Matsumoto Theorem. One checks easily that $\{a, -a\} = 0$ for any $a \in F^*$ and furthermore $\{a, b\} + \{b, a\} = 0$ for any $a, b \in F^*$ - see [B-T]. The last relation shows that the ring $K_*^M(F)$ is (graded) commutative.

For any integer m prime to the characteristic of F Kummer Theory defines a natural isomorphism $\chi : F^*/(F^*)^m \xrightarrow{\sim} H^1(F, \mu_m)$. A well-known result (apparently due to John Tate) shows that $\chi(a) \cup \chi(1-a) = 0 \in H^2(F, \mu_m^{\otimes 2})$. Thus we get an induced homomorphism of graded rings

$$\chi : K_*^M(F)/m \to \coprod_{n=0}^{\infty} H^n(F, \mu_m^{\otimes n})$$

which is known as the norm residue Homomorphism. In degrees 0 and 1 the homomorphism χ is obviously an isomorphism. One of the most interesting and nontrivial conjectures in the Galois cohomology theory of fields states that the map

Typeset by $\mathcal{A}\mathcal{M}\mathcal{S}$-TeX

173

χ is an isomorphism in all degrees (and for all m prime to *char F*). In this form this conjecture was apparently first formulated by Kazayo Kato [K], a similar (but slightly weaker) conjecture was proposed by S. Bloch [B-1]. We'll refer to the above conjecture as Bloch-Kato Conjecture. A special case of the Bloch-Kato Conjecture (for $m = 2$) was first considered by John Milnor [Mi-1] and so is often called the Milnor Conjecture. In a recent remarkable work [V 5] Vladimir Voevodsky proved the above conjecture of Milnor.

Theorem 1. [V 5] *For any field F of characteristic $\neq 2$ and any $n \geq 0$ the norm residue homomorphism*

$$K_n^M(F)/2K_n^M(F) \to H^n(F, \mathbb{Z}/2)$$

is an isomorphism.

The proof of this theorem uses essentially the motivic cohomology theory developed during the last years by V. Voevodsky in collaboration with A. Suslin and E. Friedlander - see [F-V], [S-V 1], [S-V 2], [S-V 3],[V-1], [V-2], [V-3], [V-4]. The proof also uses the Stable Homotopy Theory for schemes introduced by F. Morel and V. Voevodsky [M], [M-V]. Significant part of the latter theory is not published yet so that in a sence the proof is not quite complete.

Voevodsky works more generally with the Bloch-Kato conjecture for arbitraty m and reduces it to certain quite concrete questions concerning the universal splitting varieties. For $m = 2$ the corresponding question is known to have a positive answer due to the work of M. Rost [Ro 2]. Recently M. Rost proved that this question also has a positive answer when $m = 3$, $n = 3, 4$, thus proving that the norm residue homomorphism

$$K_n^M(F)/3K_n^M(F) \to H^n(F, \mu_3^{\otimes n})$$

is an isomorphism for $n = 3, 4$.

In preparing these notes I used heavily Bruno Kahn's report at Bourbaki seminar [K], which I found extremely helpful.

§1. THE BOLCH-KATO CONJECTURE.

In this section we make a few standard but useful general observations about the Bloch-Kato conjecture.

let F be a field of exponential characteristic p and let m be a positive integer prime to p. Consider the following statement

$BK_n(F, m)$. *The natural homomorphism*

$$\chi : K_n^M(F)/m \to H^n(F, \mu_m^{\otimes n})$$

is an isomorphism. In other words the Bloch-Kato conjecture modulo m holds for F in degree n.

The following remark is obvious from definitions.

Lemma 1.1. *Assume that $m = m_1 m_2$, where m_1 and m_2 are relatively prime. In this case the validity of $BK_n(F, m)$ is equivalent to the validity of $BK_n(F, m_1)$ and $BK_n(F, m_2)$. In particular the validity of $BK_n(F, m)$ for all m prime to p is equivalent to the validity of $BK_n(F, \ell^k)$ for all prime $\ell \neq p$ and all $k > 0$.*

¿From now on we fix a prime integer ℓ and consider only fields of characteristic $\neq \ell$. Using the transfer maps in Milnor K-theory and Galois cohomology one proves immediately the following result.

Lemma 1.2. *Let E/F be a finite field extension of degree prime to ℓ. Then the validity of $BK_n(E, \ell^k)$ implies the validity of $BK_n(F, \ell^k)$.*

The above Lemma allows us to consider only fields which have no extensions of degree prime to ℓ. In particular it suffices to consider only fields containing a primitive ℓ's root of unity. The following (well-known) fact is slightly less obvious.

Lemma 1.3. *Assume that F contains a primitive ℓ's root of unity ξ. Assume further that $BK_n(F, \ell)$ and $BK_{n-1}(F, \ell)$ hold. Then $BK_n(F, \ell^k)$ holds for any $k > 0$.*

Proof. We proceed by induction on k. The induction step is made using the diagram chase in the commutative diagram

$$
\begin{array}{ccccccc}
K_{n-1}^M(F)/\ell \otimes \mu_\ell & \longrightarrow & K_n^M(F)/\ell^{k-1} & \longrightarrow & K_n^M(F)/\ell^k & \longrightarrow & K_n^M(F)/\ell \\
\cong \downarrow & & \cong \downarrow & & \downarrow & & \cong \downarrow \\
H^{n-1}(F, \mu_\ell^{\otimes n}) & \longrightarrow & H^n(F, \mu_{\ell^{k-1}}^{\otimes n}) & \longrightarrow & H^n(F, \mu_{\ell^k}^{\otimes n}) & \longrightarrow & H^n(F, \mu_\ell^{\otimes n})
\end{array}
$$

Here the bottom row is a part of the long exact cohomology sequence, corresponding to the short exact sequence of Galois modules

$$
0 \to \mu_{\ell^{k-1}}^{\otimes n} \to \mu_{\ell^k}^{\otimes n} \to \mu_\ell^{\otimes n} \to 0,
$$

the left horizontal arrow in the top row is given by the composition

$$
K_{n-1}^M(F) \otimes \mu_\ell \to K_{n-1}^M(F) \otimes F^* \xrightarrow{\text{mult}} K_n^M(F) \to K_n^M(F)/\ell^{k-1}
$$

and the left vertical arrow coincides with the isomorphism

$$
K_{n-1}^M(F)/\ell \otimes \mu_\ell \xrightarrow{\cong} H^{n-1}(F, \mu_\ell^{\otimes n-1}) \otimes \mu_\ell = H^{n-1}(F, \mu_\ell^{\otimes n}).
$$

Lemmas 1.2 and 1.3 show that the validity of $BK_{\leq n}(F, \ell)$ for all fields F (of characteristic $\neq \ell$) implies the validity of $BK_{\leq n}(F, \ell^k)$ for all F and all $k > 0$.

Lemma 1.4. *Let F be a complete discretely valuated field with the valuation ring $\mathcal{O}$ and the residue field $\overline{F}$. Assume that $\ell \neq char\overline{F}$ and $BK_n(F,\ell)$ holds. Then $BK_n(\overline{F},\ell)$ holds as well.*

Proof. Using the fact that $\mathcal{O}$ is complete and $char\ \overline{F} \neq \ell$ one checks easily that the natural homomorphisms

$$K_n^M(\mathcal{O})/\ell \to K_n^M(\overline{F})/\ell, \quad H^n(\mathcal{O},\mu_\ell^{\otimes n}) \to H^n(\overline{F},\mu_\ell^{\otimes n})$$

are isomorphisms. This allows us to construct a commutative diagram with exact rows

$$
\begin{array}{ccccccccc}
0 & \longrightarrow & K_n^M(\overline{F})/\ell & \longrightarrow & K_n^M(F)/\ell & \xrightarrow{\partial} & K_{n-1}^M(\overline{F})/\ell & \longrightarrow & 0 \\
& & \downarrow & & \downarrow & & \downarrow & & \\
0 & \longrightarrow & H^n(\overline{F},\mu_\ell^{\otimes n}) & \longrightarrow & H^n(F,\mu_\ell^{\otimes n}) & \xrightarrow{\partial} & H^{n-1}(\overline{F},\mu_\ell^{\otimes n}) & \longrightarrow & 0
\end{array}
$$

The choice of the local parameter $\pi \in \mathcal{O}$ gives compatible splittings for the above short exact sequences and hence

$$BK_n(F,\ell) \equiv BK_n(\overline{F},\ell) + BK_{n-1}(\overline{F},\ell).$$

Lemma 1.4 allows us to reduce the general case of the Bloch-Kato conjecture modulo ℓ to the case of fields of characteristic zero.

§2. MOTIVIC COMPLEXES

In this section we fix a field F and consider the category Sm/F of smooth schemes of finite type over F. We make Sm/F into a site using one of the following three topologies: Zariski topology, Nisnevich topology or etale topology. For any $X \in Sm/F$ we denote by $L(X)$ the presheaf on the category Sm/F, given by the formula

$L(X)(Y) =$The free abelian group generated by closed integral subschemes
$\qquad\qquad Z \subset X \times Y$ finite and surjective over a component of Y

One checks easily that the pesheaf $L(X)$ is actually a sheaf in the etale topology (and a fortiori in Zariski and Nisnevich topologies as well).

Consider the standard cosimplicial object $\Delta^\bullet$ in Sm/F. For any presheaf of abelian groups $\mathcal{F}$ on Sm/F we get a simplicial presheaf $C_*(\mathcal{F})$, by setting $C_n(\mathcal{F})(U) = \mathcal{F}(U \times \Delta^n)$. We'll use the same notation $C_*(\mathcal{F})$ for the corresponding complex (of degree -1) of abelian presheaves. Usually we'll be dealing with complexes of degree $+1$, in particular, we'll reindex the complex $C_*(\mathcal{F})$ (in the standard way), by setting

$$C^i(\mathcal{F}) = C_{-i}(\mathcal{F}).$$

Recall that a presheaf $\mathcal{F} : Sm/F \to Ab$ is said to be homotopy invariant, provided that for any $U \in Sm/F$ the natural homomorphism $\mathcal{F}(U) \to \mathcal{F}(U \times \mathbb{A}^1)$ is an isomorphism. One checks easily (cf. [S-V 1] Corollary 7.5) that homology presheaves of the complex $C^*(\mathcal{F})$ are homotopy invariant.

Consider the presheaf $L((\mathbb{G}_m)^{\times n})$ (here $\mathbb{G}_m$ stands for the standard multiplicative group scheme $\mathbb{G}_m = \mathbb{A}^1 - \{0\}$) and let $\mathcal{D}_n$ be the degenerate part of this presheaf, i.e. the sum of images of homomorphisms

$$L(\mathbb{G}_m^{\times n-1}) \to L(\mathbb{G}_m^{\times n})$$

given by the embeddings of the form

$$(x_1, \ldots, x_{n-1}) \mapsto (x_1, \ldots, 1, \ldots, x_{n-1}).$$

One can verify easily that $\mathcal{D}_n$ is in fact a direct summand of $L(\mathbb{G}_m^{\times n})$. The corresponding projection $p : L(\mathbb{G}_m^{\times n}) \to \mathcal{D}_n$ is given by the formula

$$p = \sum_I (-1)^{card(I)+n-1} (p_I)_*,$$

where I runs through all proper subsets of $\{1, ..., n\}$ and $p_I : \mathbb{G}_m^{\times n} \to \mathbb{G}_m^{\times n}$ is the standard coordinate projection.

Definition 2.1. *The motivic complex $\mathbb{Z}(n)$ of weight n on Sm/F is the complex $C^*(L((\mathbb{G}_m)^{\times n})/\mathcal{D}_n)[-n]$. For a smooth scheme X over F we define its motivic cohomology groups $H^i_{\mathcal{M}}(X, \mathbb{Z}(n))$ as hypercohomology* $\mathbf{H}^i_{Zar}(X, \mathbb{Z}(n))$.

Note that $\mathbb{Z}(n)$ is a complex of sheaves with transfers in the Zariski topology (actually even in the etale topology) with homotopy invariant cohomology presheaves. The following lemma, which is a special case of results of [V 1], shows that motivic cohomology may be identified with Nisnevich hypercohomology:

$$H^i_{\mathcal{M}}(X, \mathbb{Z}(n)) = H^i_{Nis}(X, \mathbb{Z}(n)).$$

Lemma 2.2. *Let C^* be a complex (of degree $+1$) of Nisnevich sheaves with transfers with homotopy invariant cohomology presheaves.*

(1) *The cohomology sheaves (in the Zariski topology) $H^i(C^*)$ are homotopy invariant Nisnevich sheaves with transfers*
(2) *For any $X \in Sm/F$ $H^*_{Zar}(X, C^*) = H^*_{Nis}(X, C^*)$.*

For any abelian group A we use the notation $A(n)$ for the tensor product complex $\mathbb{Z}(n) \otimes A$.

Proposition 2.3.

(1) *The complex $\mathbb{Z}(0)$ is naturally quasiisomorphic to $\mathbb{Z}$.*
(2) *The complex $\mathbb{Z}(1)$ is naturally quasiisomorphic to $\mathbb{G}_m[-1]$*

(3) *The complex $\mathbb{Z}(n)$ is acyclic in degrees $> n$.*
(4) *For any $n, m > 0$ there exist natural pairings (in the derived category of bounded above complexes of Nisnevich sheaves with transfers) $\mathbb{Z}(n) \otimes^L \mathbb{Z}(m) \to \mathbb{Z}(n + m)$ which are commutative and associative.*
(5) *For any m prime to char F the complex $\mathbb{Z}/m(n)$ being considered as a complex of sheaves on the etale site is naturally quasiisomorphic to $\mu_m^{\otimes n}$.*
(6) *The n-th cohomology presheaf of $\mathbb{Z}(n)$ coincides with the sheaf $\mathcal{K}_n^M$ of Milnor K-groups. In particular $H_{\mathcal{M}}^n(\operatorname{Spec} F, \mathbb{Z}(n)) = K_n^M(F)$.*

Proof. The first and the third statements are obvious. The second follows easily from the computation of singular homology of relative curves - see [S-V 1]. Construction of the pairing $\mathbb{Z}(m) \otimes^L \mathbb{Z}(n) \to \mathbb{Z}(m + n)$ is given in [V 2]. The fifth statement follows immediately from results of [S-V 1]. The last result is proved (by an easy computation) in [S-V 3].

§3. HILBERT'S THEOREM 90

One of the main technical tools used in [M-S 1],[S 1] for the proof of the Bloch-Kato conjecture for K_2 was the following result, known as Hilbert's Theorem 90 for K_2.

Theorem 3.1 [M-S 1],[S 1]. *Let E/F be a cyclic Galois extension of prime degree ℓ. Let further σ denote a generator of $\operatorname{Gal}(E/F)$. The following sequence is exact*

$$K_2^M(E) \xrightarrow{1-\sigma} K_2(E) \xrightarrow{N_{E/F}} K_2^M(F)$$

The same result was later proved (and used as the main technical tool in the proof of the Milnor conjecture in degree 3) for K_3^M and quadratic extensions in [M-S 2], [Ro 1]. Let's consider more generally the following statement.

$HT90_n(E/F)$. *Let E/F be a cyclic Galois extension. Let further σ be a generator of the Galois group $\operatorname{Gal}(E/F)$. Then the following sequence is exact*

$$K_n^M(E) \xrightarrow{1-\sigma} K_n(E) \xrightarrow{N_{E/F}} K_n^M(F).$$

Hilbert's Theorem 90 for K_1 is due to Hilbert and is usually deduced from the following cohomological statement (also due to Hilbert): $H_{et}^1(F, \mathbb{G}_m) = 0$. Lichtenbaum proposed in [Li] the following cohomological version of Hilbert's Theorem 90 in degrees > 1.

$LiHT90_n(F, \ell)$. *Let F be a field of characteristic $\neq \ell$, then $H_{et}^{n+1}(F, \mathbb{Z}_{(\ell)}(n)) = 0$.*

The main goal of this section is to relate the above cohomological version of Hilbert's Theorem 90 with the Bloch-Kato Conjecture and with the mentioned above form of the Hilbert's Theorem 90 for K_n^M. We show also that $LiHT90$ holds for fields with divisible Milnor K-groups.

To simplify matters all fields in this section are assumed to be of characteristic zero.

We start with the following easy fact relating Zariski and etale cohomology with coefficients in $\mathbb{Q}(n)$.

Lemma 3.1 [V 1]. *Let $C^\bullet$ be a bounded above complex of etale sheaves of $\mathbb{Q}$-vector spaces with transfers. Assume that homology presheaves of $C^\bullet$ are homotopy invariant. Then for any $X \in Sm/F$ the natural homomorphisms*

$$H^*_{Zar}(X, C^\bullet) = H^*_{Nis}(X, C^\bullet) \to H^*_{et}(X, C^\bullet)$$

are isomorphisms. In particular $H^*_{Zar}(X, \mathbb{Q}(n)) \xrightarrow{\sim} H^*_{et}(X, \mathbb{Q}(n))$.

Denote by $\pi : (Sm/F)_{et} \to (Sm/F)_{Zar}$ the obvious morphism of sites.

Corollary 3.2. *The natural homomorphism of complexes of sheaves with transfers*

$$\mathbb{Q}(n) \to R\pi_*\mathbb{Q}(n)$$

is a quasiisomorphism.

The following theorem (which is the main result of [S-V 3]) is much more difficult.

Theorem 3.3.

(1) *Assume that for any field F of characteristic zero, any $i \leq n$ and any $k > 0$ the Bockstein homomorphism*

$$H^i(F, \mu_{\ell^k}^{\otimes i}) \to H^{i+1}(F, \mu_{\ell}^{\otimes i})$$

is trivial. Then $BK_{<n}(F, \ell)$ holds for any field F of characteristic zero.

(2) *Assume that $BK_{\leq n}(F, \ell)$ holds for any field of characteristic zero. Then the natural homomorphism of complexes of Zariski sheaves with transfer*

$$\mathbb{Z}/\ell^k(n) \to \tau_{\leq n}R\pi_*(\mathbb{Z}/\ell^k(n))$$

is a quasiisomorphism for all $k > 0$.

Proposition 3.4. *Assume that $LiHT90_{\leq n}(F, \ell)$ holds for all fields F of characteristic zero. Then $BK_{\leq n}(F, \ell)$ holds for any field of characteristic zero.*

Proof. Consider the following commutative diagram of complexes of etale sheaves with exact rows

$$
\begin{array}{ccccccccc}
0 & \longrightarrow & \mathbb{Z}_{(\ell)}(n) & \xrightarrow{\ell^k} & \mathbb{Z}_{(\ell)}(n) & \longrightarrow & \mathbb{Z}/\ell^k(n) & \longrightarrow & 0 \\
 & & \downarrow & & \downarrow & & =\downarrow & & \\
0 & \longrightarrow & \mu_\ell^{\otimes n} & \longrightarrow & \mu_{\ell^{k+1}}^{\otimes n} & \longrightarrow & \mu_{\ell^k}^{\otimes n} & \longrightarrow & 0
\end{array}
$$

Vanishing of $H^{n+1}_{et}(F, \mathbb{Z}_{(\ell)}(n))$ implies that the Bockstein homomorphism

$$H^n_{et}(F, \mu_{\ell^k}^{\otimes n}) \to H^{n+1}_{et}(F, \mu_\ell^{\otimes n})$$

corresponding to the bottom row is trivial. Thus it suffices to use Theorem 3.3.

Corollary 3.5. *Assume that* $LiHT90_{\leq n}(F, \ell)$ *holds for all fields* F *of character-istic zero. Then the natural homomorphisms of complexes of Zariski sheaves with transfer*

$$\mathbb{Z}_{(\ell)}(n) \to \tau_{\leq n} R\pi_*(\mathbb{Z}_{(\ell)}(n)) \to \tau_{\leq n+1} R\pi_*(\mathbb{Z}_{(\ell)}(n))$$

are quasiisomorphisms.

Proof. The cohomology sheaves of the complex $R\pi_*(\mathbb{Z}_{(\ell)}(n))$ are homotopy invari-ant Zariski sheaves with transfer. Since the $(n+1)$-st homology sheaf vanishes on fields it is equal to zero -see [V 1]. This shows that the second map is a quasi-isomorphism. The first map is a quasiisomorphism after inverting ℓ - according to Corollary 3.2 and after factoring out ℓ - according to Theorem 3.3 and Proposition 3.4 and hence is a quasiisomorphism itself.

The following result is essentially due to S. Lichtenbaum [Li].

Proposition 3.6. *Assume that* $LiHT90_{\leq n}(F, \ell)$ *holds for all fields* F *of character-istic zero. Then* $HT90_{\leq n}(E/F)$ *holds for any cyclic ℓ-primary extension of fields of characteristic zero.*

Proof. Set $G = Gal(E/F)$ and consider the following short exact sequence of Galois modules (=sheaves on the small etale site of $Spec\ F$)

$$0 \to \mathbb{Z} \to \mathbb{Z}[G] \xrightarrow{1-\sigma} \mathbb{Z}[G] \to \mathbb{Z} \to 0$$

Taking Ext to $\mathbb{Z}_{(\ell)}(n)$ we get a spectral sequence converging to zero, with the following E_1-term:

$$H^*_{et}(F, \mathbb{Z}_{(\ell)}(n)) \xleftarrow{N_{E/F}} H^*_{et}(E, \mathbb{Z}_{(\ell)}(n)) \xleftarrow{1-\sigma} H^*_{et}(E, \mathbb{Z}_{(\ell)}(n)) \leftarrow H^*_{et}(F, \mathbb{Z}_{(\ell)}(n))$$

Corollary 3.5 shows that $H^n_{et}(F, \mathbb{Z}_{(\ell)}(n)) = K^M_n(F) \otimes \mathbb{Z}_{(\ell)}$. Since $H^{n+1}_{et}(F, \mathbb{Z}_{(\ell)}(n)) = 0$ we conclude that the sequence

$$K^M_n(E) \xrightarrow{1-\sigma} K^M_n(E) \xrightarrow{N_{E/F}} K^M_n(F)$$

becomes exact after tensoring with $\mathbb{Z}_{(\ell)}$. On the other hand it's trivial to see that $Ker\ N_{E/F}/Im\ (1-\sigma)$ is killed by $[E : F] = \ell^k$.

Lemma 3.7. *Let* E/F *be a cyclic extension of degree* ℓ *of fields of characteristic zero. Assume that* F *has no extensions of degree prime to* ℓ. *Assume further that the norm map* $N_{E/F} : K^M_{n-1}(E) \to K^M_{n-1}(F)$ *is surjective and* $HT_{n-1}(E/F)$ *holds. Then* $HT90_n(E/F)$ *holds as well.*

Proof. The proof is identical to the proof of the corresponding statement for K_2 given in [S 1].

Lemma 3.8 [V-5]. *Assume that $LiHT90_{\leq n-1}(F, \ell)$ holds for any field F of charac-teristic zero. Let further F be a field of characteristic zero which has no extensions of degree prime to ℓ and let E/F be a cyclic extension of degree ℓ. Then the fol-lowing sequence (in which ϕ denotes the character corresponding to the extension E/F) is exact*

$$H^{n-1}(E, \mathbb{Z}/\ell) \xrightarrow{N_{E/F}} H^{n-1}(F, \mathbb{Z}/\ell) \xrightarrow{\cup \phi} > H^n(F, \mathbb{Z}/\ell) \to H^n(E, \mathbb{Z}/\ell)$$

We skip the proof since in the most interesting case $\ell = 2$ this statement is true (and well-known) without any assumptions on F.

Corollary 3.9. *Assume that $LiHT90_{\leq n-1}(F, \ell)$ holds for any field F of character-istic zero. Let further F be a field of characteristic zero without extensions of degree prime to ℓ and such that $K_n^M(F) = \ell K_n^M(F)$. Then the formula $K_n^M(E) = \ell K_n^M(E)$ holds also for any finite extension E of F.*

Proof. This follows immediately from Lemma 3.7, using the same argument as in the case of K_2 - see [S 1].

Theorem 3.10. *Assume that $LiHT90_{\leq n-1}(F, \ell)$ holds for any field F of charac-teristic zero. Let further F be a field of characteristic zero without extensions of degree prime to ℓ and such that $K_n^M(F) = \ell K_n^M(F)$. Then $LiHT90_n(F, \ell)$ holds.*

Proof. The group $H_{et}^{n+1}(F, \mathbb{Z}_{(\ell)}(n))$ is ℓ-torsion in view of Lemma 3.1. The exact sequence

$$H_{et}^n(F, \mu_{ell}^{\otimes n}) \to H_{et}^{n+1}(F, \mathbb{Z}_{(\ell)}(n)) \xrightarrow{\ell} H_{et}^{n+1}(F, \mathbb{Z}_{(\ell)}(n))$$

shows that it would suffice to prove that $H_{et}^n(F, \mu_{ell}^{\otimes n}) = 0$.

§4. SPLITTING VARIETIES.

As we have seen in the priveous sections to prove the general case of the Bloch-Kato conjecture modulo ℓ it suffices to consider fields of characteristic zero only. Moreover it suffices to establish for such fields the cohomological version of the Hilbert's Theorem 90. Voevodsky proves that $LiHT90_n(F, \ell)$ holds by induction on n. We assume that $LiHT90_{\leq n-1}(F, \ell)$ holds for any field F of characteristic zero and try to prove that $LiHT90_n(F, \ell)$ holds for any such F as well. One important step is already done in Theorem 3.10. There we saw that $LiHT90_n(F, \ell)$ holds provided that F has no extensions of degree prime to ℓ and $K_n^M(F)/\ell = 0$. Now the general strategy is to show that for any field F of characteristic zero there exists a field extension F'/F such that

(1) F' has no extensions of degree prime to ℓ.
(2) $K_n^M(F')/\ell = 0$
(3) The natural map $H_{et}^{n+1}(F, \mathbb{Z}_{(\ell)}(n)) \to H_{et}^{n+1}(F', \mathbb{Z}_{(\ell)}(n))$ is injective.

The main step in the construction of such an extension is to construct an exten-sion which splits a given symbol $\{a_1, ..., a_n\} \in K_n^M(F)$ and to verify the injectivity of the associated map on $H_{et}^{n+1}(-, \mathbb{Z}_{(\ell)}(n))$.

Definition 4.1. *Let* $\underline{a} = (a_1, ..., a_n)$ *be a n-tuple of elements of* F^*. *We say that an extension* K/F *is the universal splitting field for the symbol* $\{\underline{a}\} = \{a_1, ..., a_n\} \in K_n^M(F)$ *(modulo* ℓ*) provided that*

(1) $\{a_1, ..., a_n\} \in \ell \cdot K_n^M(K)$.
(2) *If* L/F *is a field extension such that* $\{a_1, ..., a_n\} \in \ell \cdot K_n^M(L)$ *then there exists an* F-*point of* K *with values in* L.

We say that a smooth projective variety X/F *is a universal splitting variety for the symbol* $\{\underline{a}\} = \{a_1, ..., a_n\} \in K_n^M(F)$ *provided that the field* $F(X)$ *is a universal splitting field for* $\{a_1, ..., a_n\}$. *In other words iff the field* $F(X)$ *splits the given symbol and moreover for any splitting field* L/F *there exists a point* $x \in X$ *and a field embedding (over* F*)* $F(x) \hookrightarrow L$.

Example 4.2. *In case* $\ell = 2$ *the construction of universal splitting varieties is quite well-known. Consider the n-fold Pfister form* $<< a_1, ..., a_n >>$ *and let* ϕ *be a subform of* $<< a_1, ..., a_n >>$ *of dimension* $> 2^{n-1}$ *(what is called in the theory of quadratic forms a Pfister neighbour) then the corresponding quadric* Q_ϕ *is the universal splitting field of the symbol* $\{a_1, ..., a_n\} \in K_n^M(F)$ *modulo 2.*

The following statements explaine the further strategy.

Lemma 4.3.

(1) *Assume that for any field* F *(of characteristic zero) and any n-symbol* $\{a_1, ..., a_n\} \in K_n^M(F)$ *there exists a splitting field* L/F *such that* $H_{et}^{n+1}(F, \mathbb{Z}_{(\ell)}(n)) \to H_{et}^{n+1}(L, \mathbb{Z}_{(\ell)}(n))$ *is injective. Then* $LiHT90_n(F, \ell)$ *holds for any* F.
(2) *Assume that there exists a splitting field* L/F *for the symbol* $\{a_1, ..., a_n\} \in K_n^M(F)$ *such that* $H_{et}^{n+1}(F, \mathbb{Z}_{(\ell)}(n)) \to H_{et}^{n+1}(L, \mathbb{Z}_{(\ell)}(n))$ *is injective. Let further* X/F *be a universal splitting variety for* $\{a_1, ..., a_n\}$, *then the natural map* $H_{et}^{n+1}(F, \mathbb{Z}_{(\ell)}(n)) \to H_{et}^{n+1}(F(X), \mathbb{Z}_{(\ell)}(n))$ *is also injective.*

The main step in Voevodsky's proof of the Milnor Conjecture is the following result.

Theorem 4.4. *Let* $\underline{a} = (a_1, ..., a_n)$ *be a n-tuple of elements of* F^*. *Denote by* $Q_{\underline{a}}$ *the projective quadrique defined by the Pfister neighbour* $<< a_1, ..., a_{n-1} >> \perp < -a_n >$. *Then the natural map* $H_{et}^{n+1}(F, \mathbb{Z}_{(\ell)}(n)) \to H_{et}^{n+1}(F(Q_{\underline{a}}), \mathbb{Z}_{(\ell)}(n))$ *is injective.*

Injectivity of the map

$$H_{et}^{n+1}(F, \mathbb{Z}_{(\ell)}(n)) \to H_{et}^{n+1}(F(Q_{\underline{a}}), \mathbb{Z}_{(\ell)}(n))$$

is deduced from vanishing of an appropriate motivic cohomology group of a certain simplicial scheme. We procede to develop the necessary technique.

For any scheme of finite type X/F denote by $\check{C}(X)$ the simplicial scheme equal in degree n to X^{n+1} and face and degeneracy operators of which are given by partial projections and partial diagonal maps. Consider the following chain of morphisms of (simplicial) schemes

$$Spec\ F(X) \to X \to \check{C}(X) \to Spec\ F$$

Lemma 4.5.

(1) *Assume that X is a smooth irreducible scheme with a rational point. Then the natural homomorphisms*

$$H^*_{Zar}(F, \mathbb{Z}(n)) \to H^*_{Zar}(\check{C}(X), \mathbb{Z}(n))$$

are isomorphisms for all n.

(2) *For any smooth irreducible scheme X the natural homomorphisms*

$$H^*_{et}(F, \mathbb{Z}(n)) \to H^*_{et}(\check{C}(X), \mathbb{Z}(n))$$

are isomorphisms.

Denote by $Co(n)$ the cone of the canonical map $\mathbb{Z}_{(\ell)}(n) \to \tau_{\leq n+1} R\pi_*(\mathbb{Z}_{(\ell)}(n))$. The Lichtenbaum conjecture predicts that this complex has to be quasiisomorphic to zero. Assuming that $LiHT90_{\leq n-1}(\ell)$ holds we conclude easily from cohomological purity for motivic cohomology the following result

Lemma 4.6. *For any smooth irreducible scheme X/F the natural restriction map $H^*_{Zar}(X, Co(n)) \to H^*_{Zar}(F(X), Co(n))$ is an isomorphism*

We say, following Saltman [Sa] that an integral scheme X/F is a rational retract iff there is a non-empty open subset $U \subset X$ which is isomorphic to a retract in an open subscheme of the affine space $\mathbb{A}^N_F$ (for some $N > 0$).

Corollary 4.7. *Let $f : Y \to X$ be a dominant morphism of smooth irreducible schemes over F. Assume that the generic fiber of f is a rational retract. Then the induced map*

$$H^*_{Zar}(X, Co(n)) \to H^*_{Zar}(Y, Co(n))$$

is an isomorphism.

Proposition 4.8. *Let X/F be a smooth irreducible scheme of finite type. Assume further that $X_{F(X)}$ is a rational retract (over $F(X)$).*

(1) *The natural homomorphisms*

$$H^*_{Zar}(\check{C}(X), Co(n)) \to H^*_{Zar}(X, Co(n)) \to H^*_{Zar}(F(X), Co(n))$$

are isomorphisms.

(2) *We have an exact sequence*

$$H^{n+1}_{Zar}(\check{C}(X), \mathbb{Z}_{(\ell)}(n)) \to H^{n+1}_{et}(F, \mathbb{Z}_{(\ell)}(n)) \to H^{n+1}_{et}(F(X), \mathbb{Z}_{(\ell)}(n))$$

Since the quadric $Q_{\underline{a}}$ from Theorem 4.4 is rational over its function field we see from Proposition 4.8 that to prove Theorem 4.4 it suffices to establish the following result

Theorem 4.9. *The motivic cohomology group*

$$H^{n+1}_{Zar}(\overset{\vee}{C}(Q_{\underline{a}}), \mathbb{Z}_{(2)}(n))$$

is trivial.

Voevodsky deduces the latter theorem from the following two results (the first of which is of topological nature whereas the second one is more arithmetical).

Theorem 4.10. *Assume that* $LiHT90_{\leq n-1}(\ell)$ *holds for all fields of characteristic zero. Let further F be a subfield of* $\mathbb{C}$ *and let* X/F *be a smooth projective variety of dimension* $d = \ell^{n-1} - 1$ *such that* $s_d(X(\mathbb{C})) \not\equiv 0 \mod \ell^2$, *where* s_d *is the Chern number associated to the d-th Newton polynom. Then there is a natural injective map*

$$H^{n+1}_{Zar}(\overset{\vee}{C}(X), \mathbb{Z}_{(\ell)}(n)) \xrightarrow{\alpha} H^{2\frac{\ell^{n-1}-1}{\ell-1}+1}(\overset{\vee}{C}(X), \mathbb{Z}_{(\ell)}(\frac{\ell^{n-1}-1}{\ell-1}+1)).$$

We sketch the proof of this theorem in the next sections.

Theorem 4.11. *Assume that* $LiHT90_{\leq n-1}(2)$ *holds for all fields of characteristic zero. Let* $Q_{\underline{a}}$ *be the quadric of Theorem 4.4. Then* $s_d(Q_{\underline{a}}(\mathbb{C})) \not\equiv 0 \mod 4$ *and*

$$H^{2^n-1}_{Zar}(\overset{\vee}{C}(Q_{\underline{a}}), \mathbb{Z}_{(2)}(2^{n-1})) = 0.$$

Computation of the Chern numbers of quadrics over $\mathbb{C}$ is well-known and easy. In particular we have $s_d(Q_{\underline{a}}(\mathbb{C})) = 2(2^{2^{n-1}-1} - 2^{n-1} - 1)$ - see [Mi-St]. The second statement of the Theorem is essentially equivalent to the Theorem of Rost [Ro 2], concerning injectivity of the norm map $A_0(X, K_1) \to F^*$ in case of Pfister quadrics. Computation of the group $A_0(X, K_1)$ for universal splitting varieties is the only part still missing in the proof of the Bloch-Kato conjecture for primes $\ell \neq 2$.

§5. Stable Homotopy Category of Schemes.

Denote by $Shv_{Nis}(Sm/F)$ the category of Nisnevich sheaves on Sm/F. Consider further the category $\Delta^{op}Shv_{Nis}(Sm/F)$ of simplicial sheaves. For any simplicial smooth scheme X_* we'll use the same notation X_* for the corresponding simplicial representable sheaf. Every scheme $X \in Sm/F$ may be considered as a constant simplicial scheme and hence also as an object of $\Delta^{op}Shv_{Nis}(Sm/F)$. A morphism of simplicial sheaves $\phi : \mathcal{X} \to \mathcal{Y}$ is said to be a weak equivalence provided that for any $u \in U \in Sm/F$ the associated map of simplicial sets $\mathcal{X}(U_u^h) \to \mathcal{Y}(U_u^h)$ is a weak equivalence. Denote by $\mathcal{H}_s(\Delta^{op}Shv_{Nis}(Sm/F))$ the localization of the category $\Delta^{op}Shv_{Nis}(Sm/F))$ with respect to the weak equivalences. An object $\mathcal{X} \in \Delta^{op}Shv_{Nis}(Sm/F))$ is said to be $\mathbb{A}^1$-local provided that for any $\mathcal{Y} \in \Delta^{op}Shv_{Nis}(Sm/F))$ the natural map $Hom_{\mathcal{H}_s}(\mathcal{Y}, \mathcal{X}) \to Hom_{\mathcal{H}_s}(\mathcal{Y} \times \mathbb{A}^1, \mathcal{X})$ is bijective . A morphism $f : \mathcal{Y} \to \mathcal{Y}'$ in $\Delta^{op}Shv_{Nis}(Sm/F))$ is said to be a weak $\mathbb{A}^1$-equivalence provided that for any $\mathbb{A}^1$-local $\mathcal{X}$ the corresponding map

$$Hom_{\mathcal{H}_s}(\mathcal{Y}', \mathcal{X}) \xrightarrow{f^*} Hom_{\mathcal{H}_s}(\mathcal{Y}, \mathcal{X})$$

is bijective. A morphism ϕ of simplicial sheaves is said to be a cofibration provided that it is injective. One checks easily (see [M-V]) that $\Delta^{op}Shv_{Nis}(Sm/F)$ with weak $\mathbb{A}^1$-equivalences and cofibrations is a closed model category in the sence of Quillen [Q]. The corresponding homotopy category $\mathcal{H}(F)$ is called the homotopy category of schemes over F. We'll use the notation $\mathcal{H}_\bullet(F)$ for the corresponding category of pointed simplicial sheaves. For any pointed simplicial sheaves $\mathcal{X}, \mathcal{Y}$ we define their smash producr as the sheafification of the presheaf $U \mapsto \mathcal{X}(U) \wedge \mathcal{Y}(U)$. This gives $\mathcal{H}_\bullet$ a structure of a simplicial monoid category.

There are two fundamental circles in the category $\mathcal{H}_\bullet$

(1) The usual simplicial circle S_s^1 considered as the constant simplicial sheaf.
(2) S_t^1 is the pointed scheme $(\mathbb{G}_m, 1)$ considered as a representable sheaf (constant in the simplicial direction).

We denote by T the pointed simplicial sheaf given by the cocartesian square

$$
\begin{array}{ccc}
\mathbb{G}_m & \longrightarrow & \mathbb{A}^1 \\
\downarrow & & \downarrow \\
Spec\ F & \longrightarrow & T.
\end{array}
$$

There are canonical isomorphisms in $\mathcal{H}_\bullet$:

$$S_s^1 \wedge S_t^1 \cong T \cong (\mathbb{P}^1, 0).$$

For any morphism $f : \mathcal{X} \to \mathcal{Y}$ of pointed simplicial sheaves one defines it's cone as the sheafification of the presheaf $U \mapsto cone(\mathcal{X}(U) \to \mathcal{Y}(U))$. Repeating this construction one gets in the usual way the cofibration sequence

$$\mathcal{X} \xrightarrow{f} \mathcal{Y} \to cone(f) \to S_s^1 \wedge \mathcal{X} \to \dots$$

Definition 5.1. *A T-spectrum over F is a family of pointed simplicial sheaves and their morphisms $\mathbf{E} = (E_i, e_i : T \wedge E_i \to E_{i+1})_{i \in \mathbb{Z}}$*

A morphism of T-spectra is defined in an obvious way. Using the above $\mathbb{A}^1$-equivalences and cofibrations one defines in the usual way weak equivalences and fibrations of spectra and finally defines the associated stable homotopy category $\mathcal{SH}(F)$. For any pointed simplicial sheaf $\mathcal{X}$ one defines its suspension spectrum in the usual way, by setting

$$\Sigma_T^\infty(\mathcal{X}) = (T^{\wedge i} \wedge \mathcal{X}, Id).$$

Abusing the language we'll use sometimes the notation $\mathcal{X}$ instead of $\Sigma_T^\infty(\mathcal{X})$.

Theorem 5.2 [V-M]. *The category $\mathcal{SH}(F)$ has a natural structure of the tensor triangulated category such that*

(1) *the shift functor $\mathbf{E} \mapsto \mathbf{E}[1]$ coincides with the smash product by S_s^1.*
(2) *The functor Σ_T^∞ takes cofibration sequences to distinguished triangles.*
(3) *The functor Σ_T^∞ takes smash products of pointed simplicial sheaves to tensor products of spectra.*
(4) *The object T of $\mathcal{SH}(F)$ is invertible.*

In the standard way each spectrum defines a cohomology theory on the category Sm/F. This theory is now bigraded (since we have two circles). More precisely to each T-spectrum $\mathbf{E}$ we associate a cohomology theory which associates to each (simplicial) smooth scheme $\mathcal{X}$ the bigraded abelian group

$$\mathbf{E}^{p,q}(\mathcal{X}) = Hom_{\mathcal{SH}(F)}(\Sigma_T^\infty(\mathcal{X}_+)), \mathbf{E}(q)[p]),$$

where $\mathbf{E}(q)[p]$ is the spectrum $\mathbf{E} \wedge S_t^q \wedge S_s^{p-q}$). We need the Eilenberg-MacLane spectra which represent the motivic cohomology theory.

For any $n > 0$ denote by $K(\mathbb{Z}(n), 2n)$ the sheaf of abelian groups $L(\mathbb{A}^n)/L(\mathbb{A}^n \setminus 0)$, considered as a pointed (by zero) simplicial sheaf. For all n there are canonical morphisms of pointed simplicial sheaves

$$e_n : T \wedge K(\mathbb{Z}(n), 2n) \to K(\mathbb{Z}(n+1), 2n+2)$$

Thus we get a spectrum $\mathbf{H}_\mathbb{Z}$ called the Eilenberg-MacLane spectrum. In the same way one defines the Eilenberg-MacLane spectra $\mathbf{H}_{\mathbb{Z}_{(\ell)}}$, $\mathbf{H}_{\mathbb{Z}/\ell}$.

Theorem 5.3 [V-M]. *For any smooth simplicial scheme $\mathcal{X}$ we have natural isomorphisms*

$$\mathbf{H}_\mathbb{Z}^{p,q}(\mathcal{X}) = H_{Zar}^p(\mathcal{X}, \mathbb{Z}(q)).$$

§6. The Steenrod Operations

Denote by $\mathcal{A}^{*,*}(F, \mathbb{Z}/\ell)$ the motivic Steenrod algebra modulo ℓ. Thus

$$\mathcal{A}^{p,q}(F, \mathbb{Z}/\ell) = Hom_{\mathcal{SH}(F)}(\mathbf{H}_{\mathbb{Z}/\ell}, \mathbf{H}_{\mathbb{Z}/\ell}(q)[p])$$

The structure of the motivic Steenrod algebra is studied in [V 4].

Theorem 6.1 [V 4].
  (1) $\mathcal{A}^{p,q}(F, \mathbb{Z}/\ell) = 0$ for $q < 0$.
  (2) $\mathcal{A}^{0,0}(F, \mathbb{Z}/\ell) = \mathbb{Z}/\ell$.

Theorem 6.2 [V 4]. *There exists a unique series of cohomology operations $P^i \in \mathcal{A}^{2i(\ell-1),\ell-1}$, $i \geq 0$, with the following properties*
  (1) $P^0 = Id$
  (2) *For any simplicial smooth scheme $\mathcal{X}$ and any $u \in H^n(\mathcal{X}, \mathbb{Z}/\ell(i))$ one has $P^i(u) = 0$ for $n < 2i$ and $P^i(u) = u^\ell$ for $n = 2i$.*
  (3)

$$\Delta(P^i) = \sum_{a+b=i} P^a \otimes P^b + \tau \sum_{a+b=i-2} \beta P^a \otimes \beta P^b$$

   where β is the Bockstein operation and τ is multiplication by $-1 \in H^0(F, \mathbb{Z}/2(1))$ if $\ell = 2$ and $\tau = 0$ otherwise.

Using the Steenrod operations P^i one defines in the standard way the Milnor operations $Q_i \in \mathcal{A}^{2\ell^i-1,\ell^i-1}$ by setting $Q_0 = \beta, Q_{i+1} = [Q_i, P^{\ell^i}]$. The same as in the usual topological situation the operations Q_i are square-zero and admit a lifting to operations $\tilde{Q}_i$ in motivic cohomology with coefficients in $\mathbb{Z}_{(\ell)}$. Define the operation

$$H^{n+1}_{Zar}(\check{C}(X), \mathbb{Z}_{(\ell)}(n)) \xrightarrow{\alpha} H^{2\frac{\ell^{n-1}-1}{\ell-1}+1}(\check{C}(X), \mathbb{Z}_{(\ell)}(\frac{\ell^{n-1}-1}{\ell-1}+1)).$$

as the composition $\tilde{Q}_{n-2}...\tilde{Q}_1$. The proof of injectivity of α (in the situation of Theorem 4.11) is based on the following result

Theorem 6.3. *In the assumptions and notations of Theorem 4.11 let $\mathcal{X} \in \mathcal{SH}(F)$ denote the fiber of $\Sigma^\infty_T(\check{C}(X)_+) \to S^0$. Then for any $i < n$ the complex*

$$... \to H^{p-2(\ell^i-1),q-\ell^i+1}(\mathcal{X}, \mathbb{Z}/\ell) \xrightarrow{Q_i} H^{p,q}(\mathcal{X}, \mathbb{Z}/\ell) \xrightarrow{Q_i} ...$$

is acyclic.

Voevodsky gives two proofs of this theorem. One of them uses essentially the algebraic cobordism spectrum and the topological realization functor. The other is more elementary and amounts to the construction of the explicit contracting homotopy operator for the above complex .

REFERENCES

[Ar]   M. Artin, *Brauer-Severi Varieties*, Lect. Notes in Math. **917** (1982), 194-210.

[B-T]   H. Bass, J. Tate, *The Milnor Ring of a Global Field*, Lect. Notes in Math. **342** (1973), 349-428.

[Bl-1] S. Bloch, *Lectures on Algebraic Cycles* (Duke Univ. Lecture Series, ed.), 1982.

[F-V]   E.Friedlander, V. Voevodsky, *Bivariant Cycle Cohomology*. In *Cycles, Transfers and Motivic Homology Theories,*, Annals of Math. Studies, Princeton Univ. Press - to appear.

[K]   B. Kahn, *La Conjecture de Milnor (d'apres V. Voevodsky)*, Seminaire Bourbaki, ex 834, (1997).

[Ka]   K. Kato, *A Generalization of Higher Class Field Theory by Using K-groups, I*, J. Fac. Sci., Univ. Tokyo **26** (1979), 303-376.

[Li]   S. Lichtenbaum, *Values of zeta-function at non-negative integers.*, Lect. Notes Math. **1068** (1983), 127 - 138..

[Me]   A.S. Merkurjev, *On the Norm Residue Homomorphism of Degree 2*, Soviet Math. Dokl. **24** (1981), 546 - 551.

[Me 2] A.S. Merkurjev, *On the Norm Residue Homomorphism for Fields*, Amer. Math. Soc. Transl. **174** (1996), 49-71.

[M-S 1] A.S. Merkurjev, A.A. Suslin, *K-cohomology of Severi-Brauer Varieties and the Norm Residue Homomorphism*, Math USSR Izv. **21** (1983), 307 - 340.

[M-S 2] A.S. Merkurjev, A.A. Suslin, *On the Norm Residue Homomorphism of Degree 3*, Math. USSR Izv. **36** (1991), 349 - 386.

[Mi 1] J. Milnor, *Algebraic K-theory and Quadratic Forms*, Invent. Math. **9** (1970), 315 - 344.

[Mi 2] J. Milnor, *An Introduction to Algebraic K-theory.* Ann. Math. Studies **76**, Princeton University Press, 1974.

[Mi-St] J. Milnor, J. Stasheff, *Characteristic Classes, Ann. Math. Studies*, vol. 76, Princeton University Press, 1974..

[M-V] F. Morel, V. Voevodsky, *Homotopy Category of Schemes over a base*, In Preparation.

[Ro 1] M. Rost, *Hilbert's Theorem 90 for K_3^M for Degree 2 Extensions*, Preprint (1986).

[Ro 2] M. Rost, *On the Spinor Norm and $A_0(X, K_1)$ for Quadrics*, Preprint (1988).

[Ro 3] M. Rost, *Chow Groups with Coefficients*, Documenta Math. 1 (1996), 319 - 393..

[Sa] D. Saltman Retract Rational Fields and Cyclic Galois Extensions, Isr. J. of Math. **47** (1984), 165 - 215..

[S-V 1] A. Suslin, V. Voevodsky, *Singular Homology of Abstract Algebraic Varieties*, Invent. Math **123** (1996), 61 - 94.

[S-V 2] A. Suslin, V. Voevodsky, *Relative Cycles and Chow Sheaves*. In *Cycles, Transfers and Motivic Homology Theories*, Annals of Math. Studies, Princeton Univ. Press.

[S-V 3] A. Suslin, V. Voevodsky, *Bloch-Kato Conjecture and Motivic Cohomology with Finite Coefficients*, Preprint (1995).

[V 1] V. Voevodsky, *Homology of Schemes*. In *Cycles, Transfers and Motivic Homology Theories*, Annals of Math. Studies, Princeton Univ. Press.

[V 2] V. Voevodsky, *Triangulated Categories of Motives over a field*. In *Cycles, Transfers and Motivic Homology Theories*, Annals of Math. Studies, Princeton Univ. Press.

[V 3] V. Voevodsky, *Bloch-Kato Conjecture for $\mathbb{Z}/2$-Coefficients and Algebraic Morava K-theories*, Preprint (1995).

[V 4] V. Voevodsky, *CohomologiCal Operations in Motivic Cohomology*, Preprint (1995).

[V 5] V. Voevodsky, *The Milnor Conjecture*, Preprint (1996).

PART II: OPEN PROBLEMS

This part of the book contains problems collected by the chairman for the corresponding area of mathematics. Any person who has an interesting open problem to propose is encouraged to contact the chairman of the appropriate division.

Open Problems in Mathematical and Applied Physics

collected by A. Jaffe and D. Stroock

Curr. Dev. Math. (1997) 193-195

Is there finite-time blow-up in 3-D Euler flow ?

U. Frisch

Observatoire de Nice
BP 4229, 06304 Nice Cedex4, France

According to Richardson's ideas on high Reynolds number turbulence, energy introduced at the scale ℓ_0, cascades down to the scale $\eta \ll \ell_0$ where it is dissipated. Consider the total time $T_\star$ which is the sum of the eddy turn-over times associated with all the intermediate steps of the cascade. From standard phenomenology à la Kolmogorov 1941 (K41), the eddy turnover time varies as $\ell^{2/3}$. If we let the viscosity ν, and thus η, tend to zero, $T_\star$ is the sum of an infinite *convergent* geometric series. Thus it takes a *finite time* for energy to cascade to infinitesimal scales. We also know that in the limit $\nu \to 0$, the enstrophy, the mean square vorticity, goes to infinity as ν^{-1} (to ensure a finite energy dissipation).

From such observations, it is tempting to conjecture that ideal flow, the solution of the (incompressible) 3-D Euler equation

$$\partial_t \mathbf{v} + \mathbf{v} \cdot \nabla \mathbf{v} = -\nabla p \qquad (1)$$
$$\nabla \cdot \mathbf{v} = 0, \qquad (2)$$

when initially regular,[1] will spontaneously develop a singularity in a finite time (finite-time blow-up).

---

[1] For example, by having only large-scale motion initially, so that the flow is very smooth, actually analytic.

This would be incorrect for at least two reasons. Firstly, the kind of phenomenology assumed above is meant only to describe the (statistically) steady state in which energy input and energy dissipation balance each other. The inviscid ($\nu = 0$) initial-value problem is not within its scope. Secondly, a basic assumption needed for K41 is that the symmetries of the Navier–Stokes or Euler equation are recovered in a statistical sense. This requires the flow to be *highly disorganized*. Kolmogorov himself was clearly aware of this, since in a footnote to his first 1941 paper he wrote:

> ... *In virtue of the chaotic[2] mechanism of translation of motion from the pulsations of lower orders to the pulsations of higher orders, ... the fine pulsations of higher orders are subjected to approximately space-isotropy statistical régime* ...

Complex spatial structures have never been observed in numerical simulations of inviscid flow with smooth initial conditions. Note that inviscid flow has frozen-in vortex lines the topology of which cannot change since no viscous reconnection can take place.

There is yet another phenomenological argument, not requiring K41, which suggests finite-time blow-up. Consider the equation for the vorticity $\omega \equiv \nabla \wedge \mathbf{v}$ for inviscid flow, written as

$$D_t\omega = \omega \cdot \nabla\mathbf{v}, \tag{3}$$

where $D_t \equiv \partial_t + \mathbf{v} \cdot \nabla$ denotes the Lagrangian derivative. Observe that $\nabla\mathbf{v}$ has the same dimensions as ω and can be related to it by an operator involving Poisson-type integrals. (For this use the fact that $\nabla^2\mathbf{v} = -\nabla \wedge \omega$.) It is then tempting to predict that the solutions of (3) will behave as the solution of the scalar nonlinear equation

$$D_t s = s^2, \tag{4}$$

which blows up in a time $1/s(0)$ when $s(0) > 0$. Actually, (4) is just the sort of equation one obtains in trying to find rigorous *upper bounds* to various norms when studying the well-posedness of the Euler problem. This is precisely why the well-posedness 'in the large' (i.e. for arbitrary $t > 0$) is an open problem in three dimensions.[3] This problem has been singled out by Saffman (1981) as 'one of the most challenging of the present time for both the mathematician and the numerical analyst'.

The evidence is that the solutions of the Euler equation behave in a way much tamer than predicted by (4). Since such evidence cannot be obtained by experimental means, one has to resort to numerical simulations. For example, Brachet *et al.* (1983) studied the Taylor-Green vortex for which

---

[2]'khaotitsheskogo' in the original

[3]In two dimensions, the absence of singularities for any $t > 0$ has been proven by Hölder (1933) and Wolibner (1933).

the initial conditions are simple trigonometric polynomials in the x, y and z variables. Using a spectral method on a 256^3 grid they found that, as long as the simulation does not run out of resolution, the width $\delta(t)$ of the complex-space analyticity strip, as a function of real time decreases exponentially. If this result can be safely extrapolated to later times, it follows that the Taylor–Green vortex will *never* develop a real-space real-time singularity: there is no inviscid blow-up. When this result was obtained in 1981, it came as a rather big surprise. Indeed, based on the kind of phenomenology described above and also on results from closure, there was a widespread belief that finite-time blow-up would take place.[4]

But is it safe to extrapolate the behavior of $\delta(t)$? About ten years later it became possible to extend the Brachet *et al.* (1983) calculation, using a grid of 864^3 points and also to study flows with random initial conditions without the somewhat special symmetries of the Taylor–Green vortex which helped in reducing computational work (Brachet, Meneguzzi, Vincent, Politano and Sulem 1992). Again, exponential decrease of $\delta(t)$ was observed.

However formidable a 864^3 simulation may look, it can only explore a span of scales of about 300, because it uses a *uniform* grid. Pumir and Siggia (1990) developed a different approach using grid-refinement 'where needed' and were thereby able to explore a span of scales of up to 10^5. Still, no blow-up was observed. Somewhat paradoxically, simulations by Grauer and Sideris (1991) and Pumir and Siggia (1992) of two-dimensional axisymmetric flow with a poloidal component of the velocity (an instance for which there is no regularity theorem) have given some evidence of finite-time blow-up. This is, however, a controversial issue (see, e.g., E and Shu 1994).

All these simulations have also given us a qualitative explanation for why ideal flow is much more regular then predicted by naive phenomenology: the exponential decrease of $\delta(t)$ corresponds to an exponential flattening of vorticity 'pancakes'. The vorticity in such structures has a very fast dependence on the spatial coordinate transverse to the pancake, so that the flow is to leading order one-dimensional. If the flow were exactly one-dimensional, the nonlinearity would vanish (as a consequence of the incompressibility condition). This *depletion of nonlinearity* explains why the growth of the vorticity is much slower than predicted by (4) which ignores this phenomenon. The problem of finite-time blow-up remains thus completely open.

The material above is mostly taken from Section 7.8 of "Turbulence, the Legacy of A.N. Kolmogorov" by U. Frisch, CUP (1995). There, the reader will find all the references and additional material on depletion of nonlinearity (in Section 9.3).

---

[4]I shared such a belief, but G.I. Taylor did not, as appears from a brief statement made to S.A. Orszag in 1969 which was communicated to me privately. As for A.N. Kolmogorov I am not aware of anything he has said on this matter.

© 1999 International Press
Curr. Dev. Math. (1997) 197 - 198

Problems

E. Lieb

Department of Mathematics
Princeton University
Princeton, NJ 08544

1. Derive a rigorous version of Quantum electrodynamics for low energies (atoms and molecules). It will involve a cutoff, but all renormalization effects should be logarithmic in the cutoff.

2. Prove long range order in the Hubbard model in three or more dimensions. (Also in two dimensions for the ground state.) (See E.H. Lieb, 'The Hubbard model – Some Rigorous Results and Open Problems', in *Proceedings of the XIth International Congress of Mathematical Physics*, Paris, 1994, D. Iagolnitzer ed., pp. 392-412, International Press (1995).

3. Forty years ago Yang and others worked out the first two terms in the ground state energy of a hard sphere Bose gas as a function of density for low density. It was non rigorous, based on Bogoliubov's theory (his seminal 1947 paper, whose existence seems to be little known nowadays). Very recently Yngvason and I have given a rigorous derivation of the first term. However, a rigorous derivation of the second term is still to be found. (If you are more ambitious, you can try to derive the third term, for which T.T. Wu gave an expression).

(see E.H. Lieb and J. Yngvason, Ground State Energy of the Low Density Bose Gas, Phys. Rev. Lett. **80**, 2504-2507 (1998).

4. Prove the existence of Bose-Einstein condensation in a continuum gas. (It is already known for a lattice model. (See T. Kennedy, E.H. Lieb and S. Shastry, "The XY Model has Long-Range Order for all Spins and all Dimensions Greater than One", Phys. Rev. Lett. **61**, 2582-2584 (1988).)

© 1999 International Press
Curr. Dev. Math. (1997) 199 – 203

Open Problems in Number Theory

Barry Mazur

Department of Mathematics
Harvard University

Number Theory has a rich assortment of ancient and modern problems, many of which have been extensively formulated and discussed in the literature. Four problems in Number Theory were contributed (to the Number Theory portion of **"Open Problems"**), each of which is either new, or else is a new suggestion regarding one of our more well known ones.

1. On Newforms, central values, and Siegel zeroes.

Let N be a larger prime, and let $H_2^+(N)$ be set of newforms of weight two for $\Gamma_0(N)$ whose L-functions have even functional equation. We (Iwaniec-Sarnak) have recently proven that as $N \to \infty$ at least 50% of the central values of $L_{(s,f)}$ for f in $H_2^+(N)$ are larger than $(log\mathrm{N})^{-2}$.

Problem A. Show that more than 50% of these central values are nonzero.

Problem B. Better still, show that more than 50% of the central values are larger than $(log\mathrm{N})^{-A}$ for some positive constant A.

One of the consequences of our recent work, is that a solution to **Problem B** implies that there are no Siegel zeros! Of course, in view of the connection via the Birch-Swinnerton Conjectures and what is known

200

about them, one can formulate these problems in terms of the ranks of $J_0(N)/\mathbf{Q}$ and their quotients.

(Submitted by Henry Iwaniec and Peter Sarnak.)

2. Intelligent counting of rational points.

Let (V, L) be a projective manifold over a number field k endowed with a metrized ample sheaf allowing one to construct the height function $h_L :$ $V(k) \to \mathbb{R}$. The naive problem of finding asymptotics for the number of k–points of bounded height has the refined version: to understand the analytic properties of the height zeta–function $Z(V, L; s) = \sum_x h_L(x)^{-s}$.

Unfortunately, it seems that if V is not homogeneous or something like that, $Z(V, L; s)$ does not have good analytic properties, unlike the more traditional Hasse–Weil–Serre– ... L–functions.

The problem I want to draw attention to is:

Find a class of good generating series counting $V(k)$.

The model I have in mind is that of Gromov–Witten invariants and quantum cohomology which provide absolutely remarkable functions counting rational curves instead of rational points.

The geometric framework for this counting is highly non–trivial and involves drastic redefinition of the naive prescription, even if one speaks about rational curves on three–dimensional quintics. The redefinition achieves the goal of moving the problem to general position in a quite sophisticated way.

An arithmetical version of this theory, if it can be built, would require a deformation theory of embedded arithmetic curves. It would be a test of maturity of Arakelov geometry, applicable to the case of manifolds with many rational points where precise analytic results are expected, as opposed to manifolds with finite number of points where qualitative results already make us happy.

(Submitted by Yuri Manin.)

References

Yu. Manin. Problems on rational points and rational curves on algebraic varieties. *Surveys in Differential Geometry, vol. 2 (1995), Int. Press, 214–245.*

M. Kontsevich, Yu. Manin. Gromov–Witten classes, quantum cohomology, and enumerative geometry. *Comm. Math. Phys., 164:3 (1994), 525–562.*

K. Behrend. Gromov–Witten invariants in algebraic geometry. *Inv. Math., 127 (1997), 601–617.*

3. Finiteness Questions

a) Concerning the Galois group of $\mathbb{Q}$

Let K be a number field contained in $\mathbb{C}$ and S a finite set of places of $\mathbb{Q}$ (possibly including archimedean places). Let $G_{K,S}$ be the Galois group of the maximal algebraic subfield of $\mathbb{C}$ which is unramified outside of S. It has been an open question for over three decades (e.g., as raised in articles of Shafarevich in *Algebraic Number Fields*, Proc. Intl. Cong. Math. Stokholm 1962) to determine whether or not $G_{K,S}$ is topologically finitely generated. It is, however, long well known that the topological group $G_{K,S}$ does have the property that the abelianization H^{ab} of any open subgroup $H \subset G_{K,S}$ of finite index is topologically finitely generated. Using the fundamental isomorphism of Class Field Theory one can even find an explicit finite system of topological generators for H^{ab}. Moreover, Leopoldt's conjecture (still, in general, outstanding) would allow us to give a simple formula for the minimal number of elements needed to generate a subgroup of H^{ab} of finite index.

In particular, in the cases where $G_{K,S}$ itself is abelian, $G_{K,S}$ *is* topologically finitely generated, although there only finitely many such examples known and in these known cases S is contained in the set of infinite places.

Problem: Can one find instances where $G_{\mathbb{Q},S}$ is abelian when S contains at least one finite prime?

Note that if S contains the prime at infinity, and at least one finite prime p, then $G_{\mathbb{Q},S}$ is non-abelian since all p-adic Galois representations attached to cuspidal newforms of level a power of p factor through $G_{\mathbb{Q},S}$.

Is $G_{\mathbb{Q},S}$ abelian for $S = \{2\}$ or $S = \{3\}$ or $S = \{5\}$? As Lenstra pointed out, one can use the reflection principle to show, it is not abelian when $S = \{37\}$.

b) Concerning points on curves

Suppose X is a curve over a field of characteristic 0 and $\pi: A \to X$ is an abelian scheme over X. Let Γ be a group of sections of π and suppose there is no element of Γ which factors through a subfamily of group varieties of relative dimension one. For $P \in X(\bar{K})$, let Γ_P be the fiber of Γ above P. Then show for all but finitely many P.

$$rank(\Gamma_P) = rank(\Gamma).$$

This statement implies both the Manin-Mumford conjecture (Raynaud's theorem) and the Mordell conjecture (Faltings' theorem). Indeed, suppose X is complete. Let J be its Jacobian, and $a: X \to J$

be an Albanese morphism, $A = X \times J$, and π the first projection. Let $M \subset J(K)$ be a subgroup of finite rank. Finally, let Γ be the group generated by the sections of $A \to X$, $\{x \mapsto (x, m) \colon m \in M\}$ and $x \mapsto (x, a(x))\}$. Then the rank of Γ_P is less than $rank(M) + 1$ only when $a(P)$ is a point in the smallest divisible subgroup of $J(K)$ group containing M. When K is a number field and M is the Mordell-Weil group $(*)$ is Mordell's conjecture and when K is algebraically closed and $M = 0$, $(*)$ is the Manin-Mumford conjecture.

c) Concerning Endomorphism Rings

Let K be a number field and g a positive integer. Then show that the number of distinct endomorphism rings of abelian varieties of dimension g defined over K is finite. This is known when $g = 1$. When $g = 2$ it implies theorems (qualitatively) of Mazur and Merel. Indeed, if E and E' are two non-CM elliptic curve there is a cyclic N-isogeny between them then the endomorphism ring of $E \times E'$ is isomorphic to $\Gamma_0(N)$. If they are not isogenous but there exists a Galois isomorphism $\phi \colon E[N] \to E'[N]$, then the endomorphism ring of the quotient of $E \times E'$ by the graph of ϕ has endomorphism ring $\{(a, b) : a, b \in \mathbb{Z}, a \equiv b \, modulo \, N\}$.

(Submitted by Robert Coleman.)

4. On the S-unit equation and polylogarithms.

If S is a finite set of rational primes, let $X(S)$ denote the set of rational numbers $x \in \mathbb{Q}$ such that both x and $1 - x$ are S-units, and denote by $N(s)$ the maximum of the cardinalities $|X(S)|$ as S ranges through all finite sets of primes with $|S| = s$. By results of Evertse, one has that $N(s) \le 1000 \cdot 50^s$, and by results of Erdös-Stewart-Tijdeman one has that $N(s)$ is bounded from below by $\exp\left((2 - o(1))\sqrt{\frac{s}{\log} s}\right)$ for large s. One expects $N(s) = \exp\left(s^{\frac{2}{3} + o(1)}\right)$. For general number fields the analogue of Evertse's result still holds (with an appropriately modified bound) but the situation for the lower bound is less clear.

The functionfield analogue of these problems has direct applications to the existence of nontrivial functional equations for polylogarithms, and deserves to be studied.

Specifically let $S = \{P_1, \ldots, P_s\}$ be a finite set of irreducible polynomials in $\mathbb{Z}[t]$ and let $X(S)$ denote the set of $x \in \mathbb{Z}[t][1/P_1, \ldots, 1/P_s]$ such that both x and $1 - x$ are units. This is a finite set $[L]$.

Problem: Is there a lower bound for $|X(S)|$ (or for the maximum of the cardinalities $|X(S)|$ ranging over sets S with $|S| = s$) which grows more than polynomially in s?

Aproof of this would imply the existence of non-trivial functional equations for polyligarithms at any level, by the result of $[Z]$, §7. Specifically, an element of the m -th Bloch group for $\mathbb{Q}(t)$ is a functional equation for $Li_m(z)$, and the number of conditions required to make an element $\sum n_i[x_i] \in \mathbb{Q}[X(S)]$ belong to the m-th Bloch group grows like s^m.

[L] S.Lang: Integral points on curves, Publ. Math. IHES **6** 1960 27-43.

[Z] D.Zagier: Polylogarithms, Dedekind zeta functions, and the algebraic K - theory of fields, in *Arithmetics Algebraic Geometry* (eds. G. van der Geer, F. Oort, J. Steenbrink), Prog. in Math. **89**, Birkhäuser, Boston 1991, 391-430.

(Submitted by Don Zagier.)

Curr. Dev. Math. (1997) 205 – 208

Open Problems in Geometry and Topology

Cliff Taubes
Department of Mathematics
Harvard University

1. Thurston's Geometrization Conjecture: Let M be a closed, connected, oriented, irreducible 3-manifold. Then precisely one of the following holds:

   a) $M = S^3/\Gamma$, where $\Gamma \subset \mathrm{Isom}(S^3)$.

   b) $\mathbb{Z} \oplus \mathbb{Z} \subset \pi_1(M)$.

   c) $M = H^3/\Gamma$, where $\Gamma \subset \mathrm{Isom}(H^3)$.

   Here, S^3 is the 3-sphere with its round metric and H^3 is the open 3-ball with its hyperbolic metric. Various portions of this conjecture are known to be true, see for example Gabai's article in *Surveys in Differential Geometry, 1996*, S. T. Yau ed., International Press, to appear.

   (This is a well known problem.)

2. Does there exist another knot in the 3-sphere with the same Jones polynomial as the unknot?

   (This is a well known problem.)

3. A generalized knot in S^3 is an oriented, immersed circle with only double point self intersections, and where the two tangent vectors at each double point are distinct. Now, let I denote and ambient, isotopy invariant of generalized knots in S^3 with at least n double points. Suppose that I integrates to an invariant of generalized knots with

$n - 1$ double points. Does I then integrate to an invariant of honest knots?

Note that an invariant, I, immersed circles with n double points is said to integrate to an invariant of circles with $n - 1$ double points when there exists and invariant of the latter, I', with the property that $I(K) = \sum_{p \in D}(I'(K_{p+}) - I'(K_{p-}))$. Here, the sum is over the set D of double points of K; and $K_{p\pm}$ are the generalized knots with $n-1$ double points which are obtained from K as follows: First, view K so that a neighborhood of p in K looks like and $\times$ with both branches oriented to point towards the top of the page. Then, K_+,is obtained by resolving the cross with the branch/passing over the $\backslash$ branch. Meanwhile, K is obtained by resolving the double point using the opposite crossing.

(Submitted by D. Bar-Natan.)

4. Is the smooth, 4-dimensional Poincare conjecture true? That is, if a compact, 4-dimensional manifold has vanishing π_1 and π_2, is it necessary diffeomorphic to S^4?

   Note that such a manifold is homeomorphic to S^4 by Freedman's theorem.

   (This is a well known problem.)

5. If $k > 2$, does every smooth, compact, $2k$-dimensional manifold which admits an almost complex structure have the structure of a complex manifold? (The almost complex structure as a complex manifold need not to be homotopic to the given one.)

   Note that when $k = 1$, the answer to the preceding is yes, and when $k = 2$, the answer is no.

   (This is one of S. T. Yau's problems from "Open Problems in Geometry", in *Differential Geometry: Partial Differential Equations on Manifolds*, S .T. Yau and R. Greene, ed., Symposia in Pure Math 154, Part 1, American Mathematical Society, Providence 1993.)

6. Let X be a smooth, projective variety and consider the Quillen-Segal group completion of the holomorphic mapping space, $\mathrm{Hol}(X, BU)^+$, where BU is viewed as a union of Grassmanninans. This space represents the "holomorphic K-theory" of X in that holomorphic maps from X to BU classify holomorphic vector bundles over X that are holomorphically embedded in trivial bundles. It is known that this

theory satisfies a kind of periodicity and sits between algebraic K-theory on the one hand and topological K-theory on the other. With the preceding understood, under what conditions (on X) is the inclusion $\text{Hol}(X, BU)^+ \to \text{Maps}(X, BU)$ a homotopy equivalence?

(Submitted by R. Cohen.)

7. The following is a question asked by John Moore about homotopy groups: Suppose that X is a simply connected, finite complex. If $\dim(\pi^*(X) \otimes \mathcal{Q})$ is finite, is there, for each prime p, and integer k for which p^k annihilates the p-torsion in $\pi^*(X)$?

This conjecture and others of Moore are surveyed in the article "Moore conjectures" by Paul Sellick in *Algebraic Topology-Rational Homotopy Theory*, Springer Lecture Notes in Math 1318, Y. Felix ed. 1988, Springer-Verlag, Berlin.

(Submitted by M. Hopkins.)

8. Does the mapping class group of a surface of genus $g \geq 2$ satisfy Kazhdan's Property T?

What follows is the definition of Property T of a finitely generated group (as is the mapping class group). Fix a finite set of generators for the group. Then, there exists a positive number ε with the following significance: Suppose that $\mathcal{H}$ is a Hilbert space on which the group has a unitary representation. If there exists a vector $v \in \mathcal{H}$ with norm 1 such that $\|\gamma - v\| < \varepsilon$ for all elements γ from the generating set, then there exists a non-zero vector in $\mathcal{H}$ which is fixed by the group.

For further reference, see "La propriete (T) de Kazhdan pour les groupes localement compacts" by P. de la Harpe and A. Valette in Asterisque **149**(1989).

(Submitted by B. Farb.)

9. Let π be a Poincare duality group. That is, the Eilenberg-MacLane space $K(\pi, 1)$ satisfies Poincare duality with respect to a fundamental class in $H^m(K; \mathbb{Z})$ for some m. Prove or disprove the "Borel conjecture": $K(\pi, 1)$ is simple homotopy equivalent to a closed, topological m-manifold which is unique up to homeomorphism.

(Unattributed.)

10. Settle the Euclidean no hair question: Does the four dimensional sphere have a unique Einstein metric (up to diffeomorphism)?

(Submitted by St. T. Yau.)

Open Problems in Analysis

collected by D. Jerison

The Cauchy problem for Nonlinear Schrödinger Equation (NLS) with critical nonlinearity

J. Bourgain
IAS at Princeton

Consider the initial value problem

$$\begin{cases} iu_t - \Delta u + u|u|^{p-2} = 0 \\ u(0) = \phi \in H^s(\mathbb{R}^D) \end{cases} \tag{1}$$

corresponding to defocusing NLS with Hamiltonian

$$H(\phi) = \int |\nabla \phi|^2 + \int |\phi|^p. \tag{2}$$

If we fix $s \geq 0$, then the critical nonlinearity is given by

$$p - 2 = \frac{4}{D - 2s}. \tag{3}$$

In this case, there is local wellposedness on a time interval $[0, T^*[$ and global wellposedness $(T^* = \infty)$ for small data (these facts hold also in the focusing case). The question is whether in the defocusing case, the smallness assumption may be dropped in the global existence result. Two cases are particularly significant because of conservation laws.

(I) <u>$s = 0$</u>: The L^2-critical case

Global wellposedness holds if $\phi \in H^1$ or $(1 + |x|)\phi \in L^2$.

The question remains for $\phi \in L^2$

(II) $\underline{s = 1}$: The H^1-critical case.

In particular, for 3D, the global existence of classical solutions for

$$iu_t - \Delta u + u|u|^4 = 0 \qquad (4)$$

is open (one also expects scattering).

This is perhaps the most important question in this area.

Comments

Questions (I), (II) are related. In both cases, the issue consists in disproving certain concentrations (of L^2 or H^1-norm) on small regions in space-time by means of apriori inequalities. For the 3D wave equation

$$y_{tt} - \Delta y + y^5 = 0 \qquad (5)$$

the problem was solved by Struwe (radial case) [S] and Grillakis [G]. It is basically an interplay between Strichartz's theory and the apriori inequality of Morawetz. On an heuristic level, the main difference between (9) and (5) is infinite propagation speed. The question is open for (4) also in the radial case. In fact, no global solutions that are not small in some sense seem known. See [G-V], [L-S], [C] for the NLS background.

REFERENCES

[S]. M. Struwe, *Globally regular solutions to the u^5-Klein-Gordon equations*, Ann. Scuola Norm. Sup. Posa. Ser 5, 15 (1988), 495-513.

[G]. M. Grillakis, *Regularity and asymptotic behaviour of the wave equation with classical nonlinearity*, Ann. of Math. 132 (1990), 485-509.

[G-V]. J. Ginibre, G. Velo, *Scattering theory in the energy space for a class of nonlinear Schrödinger equations*, J. Math. Pures et Appl. 64 (1985), 363-401.

[L-S]. J. Lin, W. Strauss, *Decay and scattering of solutions of a nonlinear Schrödinger equation*, J. Funct. Anal. 30 (1978), 245-263.

[C]. T. Cazenave, *An introduction to nonlinear Schrödinger equations*, Textos de Metodos Matematicos 26, (Rio de Janeiro)..

Singularities of Schrödinger's equation and recurrent bicharacteristic flow

Walter Craig
Mathematics Department
Brown University
Providence, Rhode Island 02912

This note is concerned with solutions of the time dependent Schrödinger equation

$$(1) \qquad i\partial_t \psi = -\tfrac{1}{2} \sum_{j,\ell=1}^{n} \partial_{x_j} a^{j\ell}(x) \partial_{x_\ell} \psi,$$

with $x \in \mathbf{R}^n$. The Schrödinger kernel $S(x, y, t)$ is a distribution on $\mathbf{R}_x^n \times \mathbf{R}_y^n \times \mathbf{R}_t$, and we are interested in a microlocal description of its singularities. It is usual to consider modifications of equation (1) with an additional potential term or with first order terms which arise when describing the effects of an external potential or an electro-magnetic field. The situation for (1) is however already interesting, and for the most part we will focus on it. Much of the analysis carries through in the more general setting of a complete Riemannian manifold whose non-compact ends are endowed with scattering metrics, again we will mostly remain with the case at hand. The discussion below consists of excerpts from two seminar talks given at the MSRI in Berkeley on 23 October and the Fields Institute in Toronto on 28 October 1997.

213

The free euclidian case corresponds to $a^{j\ell} = \delta^{j\ell}$, for which

$$(2) \qquad\qquad i\partial_t \psi = -\tfrac{1}{2}\Delta\psi \ ,$$

and the Schrödinger kernel is explicitely

$$(3) \qquad\qquad S^0(x - y, t) = \frac{1}{\sqrt{2\pi it}^n} e^{\frac{i|x-y|^2}{2t}} \ .$$

From this expression we observe a number of self evident facts.

Proposition 1: *Given $\psi(x,t)$ a solution of (2) with $\psi_0(x) \in L^2(\mathbf{R}^n)$, define the probability distributions*

$$dP_t(x) = |\psi(x,t)|^2 dx \ .$$

(i) Schrödinger evolution conserves probability,

$$\int_{\mathbf{R}^n} dP_t(x) = \int_{\mathbf{R}^n} dP_0(x) \ .$$

(ii) If $dP_0(x)$ possesses all of its moments,

$$(4) \qquad\qquad \int_{\mathbf{R}^n} |x^k \psi_0(x)|^2 dx < +\infty \ ,$$

then for all $t \neq 0$, the solution $\psi(x,t)$ is C^∞.
(iii) Given a distribution $\psi_0 \in \mathcal{D}'$ for initial data which possesses all of its moments, then the solution $\psi(x,t)$ is again C^∞ for all $t \neq 0$.

An interpretation of this is that all singularities in the data are carried to infinity instantly by the solution operator for equation (2). To understand the analogous phenomenon in the more general setting of equation (1), consider the principal symbol $a(x,\xi) = \tfrac{1}{2}\sum_{j,\ell} a^{j\ell}(x)\xi_j\xi_\ell$, whose bicharacteristic flow on $T^*(\mathbf{R}^n)$ we denote by $\varphi_s(x,\xi)$. The first connection between singularities and the bicharacteristic flow was given by L. Boutet de Monvel [1] and R. Lascar [8], who proved that for solutions of (1), the (appropriate quasi-homogeneous) wave front set $WF_Q(\psi(x,t)) \subseteq T^*(\mathbf{R}_x^n \times \mathbf{R}_t)$ is invariant under the flow $\Phi_s(x,\xi,t,\tau) = (\varphi_s(x,\xi),t,\tau)$. The proof of this is closely related to L. Hörmander's classical results [6] on the propagation of singularities for hyperbolic equations. In our case (1) the implication is that singularities must propagate with infinite velocity along bicharacter-istics, and that the wave front set of solutions is a 'horizontal' set in space-time. These results do not however predict the evolution of singularities in time t. For illustration, the wave front set of the free Schrödinger kernel $S^0(x,t)$ is precisely $WF(S^0(x,t)) = \{(x,\xi,t,\tau) : t = \tau = 0, \xi/|\xi| = \pm x/|x|\}$, the set of radial lines eminating from and returning to the origin at $t = 0$.

An improvement to this result can be made under a hypothesis of non-recurrence of the bicharacteristic flow. We consider elliptic operators $\sum_{j\ell} \partial_{x_j} a^{j\ell}(x)\partial_{x_\ell}$ such that

$$|\partial_x^\alpha(a^{j\ell}(x) - \delta^{j\ell})| \le \frac{C_\alpha}{\langle x \rangle^{|\alpha|+1+\varepsilon}} .$$

In case we have normalized $\det(a) = 1$ these are Laplace-Beltrami operators for the asymptotically flat metric $ds^2 = \sum_{j\ell} a_{j\ell}(x)dx^j dx^\ell$.

Theorem 2: *(W. Craig, T. Kappeler & W. Strauss [3]) Suppose that $(x_0, \xi^0) \in T^*(\mathbf{R}^n)$ is not trapped backwards, that is, it is carried to infinity by the bicharacteristic flow $\varphi(s; x, \xi)$ as $s \to -\infty$. Given data $\psi_0(x)$ which satisfies the moment condition (4), then for all $t > 0$,*

(5) $(x_0, \xi^0) \notin WF(\psi(x,t)) .$

In the nonelliptic case there are results of a similar nature which ap-pear in [3], and there are further results in recent unpublished work by C. Kenig, G. Ponce, C. Rolvung, and L. Vega. The moment con-dition is of course overly strong, one may replace it with a condition on the moments of ψ_0 appropriately microlocalized near the past of the bicharacteristic through (x_0, ξ^0).

A number of global corollaries follow directly from this statement. Define the *recurrent set* of the bicharacteristic flow to be $R = \{(x, \xi) \in T^*(\mathbf{R}^n)\backslash\{0\} : \varphi(s : x, \xi) \text{ is bounded for all } s \in \mathbf{R}\}$, and let $M^u(R) = \{(x, \xi) : \varphi(s : x, \xi) \text{ is bounded for all } s \le 0\}\backslash R$.

Corollary 3: *(i) Whenever $\psi_0(x)$ satisfies (4), then for all $t > 0$*

$$WF(\psi(x,t)) \subseteq R \cup M^u(R) .$$

(ii)   Consider the Schrödinger kernel $S(x, y, t)$.   Whenever either (x_0, ξ^0) is not trapped backwards, or else (y_0, η^0) is not trapped forwards by the bicharacteristic flow $\varphi(s; x, \xi)$, then for all $t > 0$

$$(6)\qquad\qquad (x_0, \xi^0, y_0, \eta^0) \notin WF(S(x, y, t)) .$$

Recently J. Wunsch [9] has substantially improved this theorem and its corollaries in the setting of scattering metrics at infinity. However it is evident that the picture of the singularities of the Schrödinger kernel is far from complete, and that it is intimately connected with the recurrence properties of the flow $\varphi(s)$. Because of infinite propagation speed, it is not sufficient to connect a point $(y_0, \eta^0) \in T^*(\mathbf{R}^n)$ with $(x_0, \xi^0) \in T^*(\mathbf{R}^n)$ with a bicharacteristic, in order that $(x_0, \xi^0, y_0, \eta^0) \in WF(S)$. Rather we must have a situation in which the bicharacteristic flow of neighborhoods of (y_0, η^0) is recurrent at (x_0, ξ^0). The following conjecture is therefore compelling.

Conjecture 4: *For $t > 0$ the point $(x_0, \xi^0, y_0, \eta^0)$ is in $WF(S(x, y, t))$ only if, for any conic neighborhoods Ω_1, Ω_2, with $(x_0, \xi^0) \in \Omega_1$ and $(y_0, \eta^0) \in \Omega_2$, then there is a sequence of times $s_j \to +\infty$ such that*

$$\varphi(s_j; \Omega_2) \cap \Omega_1 \neq \emptyset .$$

Aside from several examples which are compatible with this statement and a few results in special cases, this question remains open.

The simplest example of a recurrent situation is to pose equation (2) in one dimension on the circle S^1. The Schrödinger kernel is again explicit

$$(7)\qquad\qquad S^{per}(x - y, t) = \sum_{k \in \mathbf{Z}} \frac{1}{2\pi} e^{ik^2 t/2} e^{ik(x-y)} ,$$

yet it is nonetheless interesting. At rational times $t/2\pi = p/q$ (in lowest terms) S^{per} decomposes into a linear combination of δ-functions, supported at translates of the qth roots of unity. Note that $WF(S^{per}) \subset T^*(S_x^1) \times T^*(S_y^1)$ is thus intermittent, while $WF_Q(S^{per}) \subseteq T^*(S_x^1 \times S_y^1 \times \mathbf{R}_t)$ is not, and is larger. For $2\pi t \in \mathbf{R} \backslash \mathbf{Q}$ the regularity of S^{per} is worse; it is not even a measure [5], and in fact its regularity is related to the diophantine properties of $t/2\pi$ [7]. In any case, all

$(y, \eta) \in T^*(S^1)$ are recurrent to all $(x, \xi) \in T^*(S^1)$, supporting the above conjecture.

A next most simple example is a manifold which is asymptotically hyperbolic, constructed from the upper half space, $M = \mathbf{H}^d \backslash \Gamma$, where Γ is generated by one hyperbolic element. When the dimension d is odd the heat kernel $H(x, y, t)$ has an explicit description in geometrical terms, and one may simply continue it for complex time to obtain the Schrödinger kernel $S(x, y, t) = H(x, y, it)$. The manifold M has one simple closed geodesic. Inspecting $S(x, y, t)$ on this periodic geodesic it is easy to see that it possesses a singularity, indeed the same singularity as for the circle S^1 of the previous example.

As a final remark, there is a theorem of S. Doi [4] which addresses the recurrence of bicharacteristics and relates it to the local smoothing property of solutions of Schrödinger's equation. His statement is essentially that a microlocal smoothing property of order $1/2$ holds on bicharacteristics which are not trapped, however it cannot hold for any conic neighborhood Ω of a point (x_0, ξ^0) at which the bicharacteristic flow is recurrent.

Acknowledgements: Research supported in part by the grant number NSF DMS-9706273.

References:

[1] L. Boutet de Monvel, 'Propagation des singularités des solutions d'équations analogues à l'équation de Schrödinger', *Colloque Int'l Univ. Nice, Lecture Notes in Math.* **459** Springer Verlag, New York Heidelberg Berlin (1974) 1–14.

[2] W. Craig, 'Les moments microlocaux et la régularité des solutions de l'équation de Schrödinger', *Séminaire Equations aux Dérivées Partielles (Mathématique, Ecole Polytechnique)* 1995-1996.
 - (english version) 'Microlocal moments and regularity of solutions of Schrödinger's equation', Math. Physics Electronic Journal 97-2 (1997), mpej@math.utexas.edu .

[3] W. Craig, T. Kappeler and W. Strauss, 'Microlocal dispersive smoothing for the Schrödin-ger equation', *Commun. Pure Applied Math.* **48** (1995) 769–860.

[4] S. Doi, 'Smoothing effects of Schrödinger evolution groups on Riemannian manifolds', *Duke Math. J.* **82** (1996) 679–706.

[5] P. Gérard, F. Golse and J. Rauch, personal communication.

[6] L. Hörmander, 'On the existence and regularity of solutions of linear pseudo - differential equations', *Enseignement Mathematiques* **17** (1971) 99–163.

[7] L. Kapitanski and I. Rodnianski, preprint.

[8] R. Lascar, 'Propagation des singularités des solutions d'équations pseudodifferentielles quasi - homogènes', *Annales Inst. Fourier* **27.2** (1977) 79–123.

[9] J. Wunsch, 'Propagation of singularities and growth for Schrödinger operators', Ph.D. thesis Harvard Univ. 1998, preprint.

Global minimizers of the Mumford-Shah functional

Guy David
Université de Paris-Sud Orsay

The Mumford-Shah functional was introduced in the late eighties as a tool for (automatic) image segmentation [MuSh]. Let us give its definition in any dimension n, even though $n = 2$ in the initial setting. Let Ω be a simple bounded domain in $\mathbb{R}^n$, like a ball or a rectangle, and let $g \in L^\infty(\Omega)$ be given. In the context of image processing, $n = 2$, Ω is a screen, g is an image, and $g(x)$ represents the level of brightness (or colour if g is vector-valued) of the image at the point x. The functional is given by

$$(1) \quad J(u, K) = \int\int_{\Omega \backslash K} |u - g|^2 + \int\int_{\Omega \backslash K} |\nabla u|^2 + H^{n-1}(K) ,$$

where the authorized competitors are pairs (u, K) such that K is a closed subset of Ω with finite Hausdorff measure $H^{n-1}(K)$ and u is a function on $\Omega \backslash K$ with one derivative ∇u in $L^2(\Omega \backslash K)$. One hopes that minimizers (u, K) of J will realize a good compromise between the three competing constraints that u be close to the original image g, u be fairly regular except on a singular set K where it may have jumps, and K itself be reasonably short. Of course the three terms of the functional should be given different weights that reflect the relative importance of the three constraints, but for theoretical mathematics the constants can be normalized out by multiplying g and u by the same constant and dilating the picture. The point here is not to have a

219

very good accuracy (as in an image compression problem), but rather
to get a fairly simple, cartoon-like description by (u, K) that would
capture significant features of the image g.

Discrete variants of the functional J are used in practice for image segmentation; we refer to [MoSo] for an interesting description of
some of them and further information on computer vision and image
segmentation. When $n = 3$, the Mumford-Shah functional has been
proposed as a model for the formation of cracks in a material, but I
don't know how useful it is in this context.

Let us now go rapidly through some of the known properties of
minimizers of J. The existence of minimizers (u, K) (for all choices
of g) was not obvious, but is known ([DCL] and [Am]). (The trouble
is the lack of semicontinuity of Hausdorff measure: it is easy to find
sequences of compact sets K_n such that $H^1(K_n)$ tends to 0 but which
converge to a line segment, say.) The main mathematical questions
concern the regularity of K for minimizers (u, K) of J. Observe that
it is a good thing for image processing if K is not too wild, because
we think of it as edges in the image.

The focus is on K because u is determined by K (when K is
fixed, minimizing J in u is a well-posed convex variational problem).
In order to discuss regularity properties of K, we should restrict to
"irreducible minimizers", i.e., minimizers (u, K) for which u does not
extend to a function with a derivative in $L^2(\Omega \backslash K')$ for any K' strictly
smaller than K. Otherwise we may artificially add any horrible com-
pact set of H^1-measure 0 to K and ruin any nice description of K

The initial conjecture of Mumford and Shah [MuSh] is that when
(u, K) is an irreducible minimizer, K is a finite union of C^1-arcs that
may meet at their endpoints with 120° angles. This is still an open
question, but significant progress has been made. Here are a few
properties of K when (u, K) is an irreducible minimizer. First, K
is rectifiable [DCL], which means that it is contained in a countable
union of C^1-curves, plus possibly a set of H^1-measure zero. Also, it
is Ahlfors-regular [DMS], which means that

$$(2) \quad C^{-1}r \ \leq \ H^1(K \cap B(x, r)) \ \leq \ Cr \text{ for all } x \in K \text{ and } 0 < r < 1.$$

Next, K is uniformly rectifiable: it is contained in an Ahlfors-
regular curve (i.e., a compact curve that satisfies (2)) [DaSe1]. Finally,
the best result of this type is that for every disk $B(x, r)$ centered on
K and radius ≤ 1 there is another disk $B(y, t) \in B(x, r)$ centered

on K, with $t \geq C^{-1}r$, and such that $K \cap B(y,t)$ is a C^1-curve. (See [Da] or [AFP]). These results have higher-dimensional analogues: see [CaLe1,2], [DaSe3], [AFP] respectively. The reader may consult [DaSe2] for a survey of some proofs, and [MoSo] for a more extensive study and other results that I did not quote here.

I wish to insist more on a recent approach introduced by A. Bonnet [Bo1] in dimension $n = 2$. He first proves that the objects obtained by blowing up a reduced minimizer (u, K) of J are "global minimizers", as defined below. Blowing up here consists in taking a point $x_0 \in K$ and a sequence of numbers t_n tending to 0, and then considering the sets $K_n = x_0 + t_n^{-1}(K - x_0)$, and suitably normalized associated functions u_n . Limits exist (modulo taking subsequences), and are pairs (v, E) with the following properties.

First, E is a closed set in the plane, and v is a function on $\mathbb{R}^2 \backslash E$ with a derivative in $L^2_{loc}(\mathbb{R}^2 \backslash E)$ such that

(3)
$$H^1(E \cap B(0,R)) < \infty \text{ and } \int_{\mathbb{R}^2 \cap B(0,R)) \backslash E} |\nabla v|^2 < \infty \text{ for all } R > 0.$$

Competitors of the pair (v, E) are similar pairs (v^*, E^*) , where E^* is also a closed set in the plane, v^* a function on $\mathbb{R}^2 \backslash E^*$, the analogue of (3) holds, and there is a radius $R > 0$ such that $v^* = v$ and $E^* = E$ out of $B(0, R)$. We also add the topological constraint that for R large enough, E^* separates all pairs of points $y, z \in \mathbb{R}^2 \backslash (E \cup B(0, R))$ that are separated by E (that is, such that y and z lie in different connected components of $\mathbb{R}^2 \backslash E$. The pair (v, E) is called a global (Mumford-Shah) minimizer if for all competitors (v^*, E^*) of (v, E) we have that

$$H^1(E \cap B(0,R)) + \int_{\mathbb{R}^2 \cap B(0,R)) \backslash E} |\nabla v|^2$$

(4)
$$\leq H^1(E^* \cap B(0,R)) + \int_{\mathbb{R}^2 \cap B(0,R)) \backslash E^*} |\nabla v^*|^2$$

for R large enough.

Note that global minimizers can be thought of as solution of some Dirichlet problem at infinity (E and the values of v are known at infinity, and we only compare with compact perturbations). The following is a list of tentative global minimizers:

1. $E = \emptyset$ and v is a constant;

2. E is a line and v is constant on each of the two components of $\mathbb{R}^2 \backslash E$;

3. E is the union of three half lines with the same endpoint and that make $120°$ angles, and v is constant on each of the three components of $\mathbb{R}^2 \backslash E$;

4. Up to Euclidean motion, E is the half line $E = \{y = 0 \text{ and } x \leq 0\}$ and

$$v(x,y) \quad = \quad \pm\sqrt{\tfrac{2}{\pi}} \quad Im\{(x \ + \ iy)^{1/2}\} \ + \ C \quad .$$

The three first examples are easily seen to be global minimizers. Note that the topological constraint on competitors is useful here. Whether the fourth one is also a global minimizer is still an open question (of De Giorgi), although A. Bonnet may have a positive answer [Bo2]. If it were not, then it would not be possible to find isolated, compactly supported line segments or open C^1-curves in any minimizer of the (local) Mumford-Shah functional J above.

The main question, though, is whether all the global minimizers are in the list above. This would imply the Mumford-Shah conjecture above (and it is probably the best way to prove it).

A. Bonnet was only able to prove this for global minimizers such that E is connected, and because of this he was "only" able to show that if (u, K) is a reduced minimizer of the Mumford-Shah functional J, every **isolated** connected component of K is a finite union of C^1-curves that meet with $120°$ angles.

Let me finish this description with a formula that should be useful. It is a (surprisingly simple) Euler-Lagrange formula that was discovered by J.-C. Léger. Identify $\mathbb{R}^2$ with the complex plane, and set

$$F(x + iy) = \frac{\partial v}{\partial z}(x + iy) = \frac{1}{2}[\frac{\partial v}{\partial x}(x + iy) - i\frac{\partial v}{\partial y}(x + iy)]$$

(5)						for $(x, y) \in \mathbb{R}^2 \backslash E$.

Note that v is harmonic because it minimizes the integral of $|\nabla v^2|$ locally, and so F is holomorphic on $\mathbb{C}\backslash E$. The formula is

(6)				$$F(z)^2 \ = \ \frac{-1}{2\pi} \int_E \frac{dH^1(w)}{(z - w)^2} \quad \text{for } z \in \mathbb{C}\backslash E \ .$$

It has a few applications, like the fact that v is determined by E modulo an additive constant and a change of sign on each component (which was not clear a priori), or the fact that all global minimizers for which E is contained in a line (or a sufficiently flat chord-arc curve) are in the list above, but so far it does not have as striking applications as I think it should.

One can also start to dream about higher dimensions. It would be nice to have a list of tentative global minimizers in dimension 3, even if the chances of proving that it is the right list are quite small at this stage. Maybe interesting new phenomena will show up.

REFERENCES

[Am]   L. Ambrosio, Existence theory for a new class of variational problems, Arch. Rational Mech. Anal. 111 (1990), 291-322.

[AFP]   L. Ambrosio, N. Fusco and D. Pallara, Partial regularity of free discontinuity sets II., preprint, Univ. de Pisa 1994.

[Bo1]   A. Bonnet, On the regularity of edges in image segmentation, Ann. Inst. H. Poincaré, Analyse non linéaire, Vol 13, 4 (1996), 485-528.

[Bo2]   A. Bonnet, Personal communication, 1996, to be retrieved.

[CaLe1]   M. Carriero and A. Leaci, Existence theorem for a minimum problem with free discontinuity set , Nonlinear Analysis, Theory, Methods and applications, Vol 15, 1990, 661-677.

[CaLe2]   M. Carriero and A. Leaci, S^k-valued maps minimizing the L^p norm of the gradient with free discontinuities, Ann. Sc. Norm. Sup. Pisa, Vol. 18, 1991, 321-352.

[DMS]   G. Dal Maso, J.-M. Morel, and S. Solimini, A variational method in image segmentation: Existence and approximation results, Acta Math. 168 (1992), 89-151.

[Da]   G. David, C^1-arcs for minimizers of the Mumford-Shah functional, SIAM Journal of Appl. Math., Vol. 56, 3, 1996, 783-888.

[DaSe1]   G. David and S. Semmes, On the singular sets of minimizers of the Mumford-Shah functional, Journal de Math. Pures et Appl. 75 (1996), 299-342.

[DaSe2]   G. David and S. Semmes, On a variational problem from image processing, proceedings of the conference in honor of J.-P. Kahane, special issue of the Journal of Fourier Analysis and Applications, 1995, 161-187.

[DaSe3]   G. David and S. Semmes, Uniform rectifiability and Singular sets, Annales de l'Inst. Henri Poincaré, Analyse non linéaire, Vol 13, 4 (1996), p. 383-443.

[DCL]   E. De Giorgi, M. Carriero, and A. Leaci, Existence theorem for a minimum problem with free discontinuity set, Arch. Rational Mech. Anal. 108 (1989), 195-218.

[DiKo]   F. Dibos and G. Koepfler, Propriété de régularité des contours d'une image segmentée, Comptes Rendus Acad. Sc. Paris 313 (1991), 573-578.

[MoSo]   J.-M. Morel and S. Solimini, Variational methods in image segmentation, Progress in nonlinear differential equations and their applications 14, Birkhauser 1995.

[MuS]   D. Mumford and J. Shah, Optimal approximations by piece-wise smooth functions and associated variational problems, Comm. Pure Appl. Math. 42 (1989), 577-685.

Guy David,
Université de Paris-Sud Orsay,
Institut Universitaire de France,
and MSRI
guy.david@math.u-psud.fr
gdavid@msri.org

The modified Bers conjecture

L. Escauriaza

Universidad del Pais Vasco
mtpeszul@lg.ehu.es

It is well known that on any 2-dimensional Lipschitz domain D, there are no non-constant harmonic functions whose gradient vanishes on a set with positive arc-length measure contained in the boundary of D. On the contrary, when the dimension is greater than 2, these bad harmonic functions do exist. In fact, in [3] the authors constructed a non-zero harmonic function u defined in the upper closed space $\mathbb{R}^3_+$, verifying that both u and its gradient vanish simultaneously on a set with positive 2-dimensional Lebesgue measure contained in the boundary of $\mathbb{R}^3_+$. This counterexample disproved the so called "Bers' conjecture", which claimed that these bad harmonic functions could not exist.

On the other hand, the following "modified Bers conjecture" still remains an open question:

Let u be a harmonic function in a connected Lipschitz domain D in $\mathbb{R}^d$, $d \geq 3$, containing the origin as a boundary point. Assume that u vanishes continuously on $B_{2R} \cap \partial D$ for some positive R. Then, u must be identically zero in D if the surface measure of the set

$$E = \{Q \in \partial D \cap B_R : \nabla u(Q) = 0\}$$

is positive.

Observe that in this case, one assumes that the tangential derivatives of u are identically zero on an open set of the boundary, and one is asking what can happen with the size of the set where the normal derivative vanishes within the open set.

This conjecture is known to hold when the boundary of D can be written locally as the graph of functions $\varphi : \mathbb{R}^{d-1} \longrightarrow \mathbb{R}$ verifying that for some $M < \infty$, $0 < \alpha \leq 1$ and all x,y in $\mathbb{R}^{d-1}$

$$(x - y) \cdot \nabla\varphi(x) - (\varphi(x) - \varphi(y)) \geq -M|x - y|^{1+\alpha}.$$

In particular, this property holds when the functions φ are the sum of a $C^{1,\alpha}$ function and a convex function [1],[2], [8]. Moreover, when D is a $C^{1,\alpha}$ domain and u is a nonzero harmonic function verifying the above conditions, it has been shown that the Hausdorff dimension of the set E is not greater than $d-2$ [9], [2]. But nothing is known about this problem when D is just a C^1 domain. This is an interesting and challenging problem in harmonic analysis, and a positive answer to the modified conjecture would have some interesting applications to control theory [7]. For instance, it would be possible to yield exact controllability for the heat operator on cylindrical domains with minimal smoothness conditions on the base of the cylinder. As shown by Lin in [9], uniqueness for boundary problems can also be used to obtain estimates for nodal sets of eigenfunctions due to Donnelly and Fefferman [4].

The theorems that have been proved so far rely either on a monotonicity method introduced by Garofalo and Lin [5] or Carleman inequalities. An exposition of the Carleman method and related conjectures can be found in [6].

References

[1]   Adolfsson A., Escauriaza E., Kenig C.E., Convex domains and unique continuation at the boundary , Rev. Mat. Iberoamericana 11,3 (1995),513-525.

[2]   Adolfsson A., Escauriaza E., $C^{1,\alpha}$ domains and unique continuation at the boundary , Comm. Pure. Appl. Math. Vol. L (1997),935-969.

[3]   Bourgain J., Wolff T., A remark on gradients of harmonic functions in dimension $d \geq 3$, Colloquium Mathematicum LX/LXI (1990), 253-260.

[4]   Donnelly H. and Fefferman C., Nodal sets of eigenfunctions on Riemannian manifolds , Inv. Math. 93 (1988), 161-183.

[5]   Garofalo N., Lin F.H., Monotonicity properties of variational integrals, A_p-weights and unique continuation , Indiana Univ. Math J. 35 (1986), 245-267.

[6]   Jerison D. and Lebeau G. Nodal sets of sums of eigenfunctions , to appear in "Harmonic Analysis and Partial Differential Equations: Essays in Honor of Alberto Calderón," ed. Michael Christ, Carlos Kenig, and Cora Sadosky, Chicago U. Press, 1999.

[7]   Kenig C.E., Spring School 94: Potential theory and analysis , Paskey nad Jizerou (Czech Republic). Lectures by Carlos E. Kenig. Notes by Michael Korey.

[8]   Kukavica I. and Nyström K., Unique continuation on the boundary for Dini domains , to appear Proc. A. M. S.

[9]   Lin F.-H., Nodal sets of solutions of elliptic and parabolic equations , Comm. Pure and Appl. Math. XLIV (1991), 287-308.

THE WAVE MAP PROBLEM

Manoussos G. Grillakis
Dept. of Mathematics
University of Maryland
College Park MD 20742

Introduction :

The simplest way to introduce the problem is by considering an unknown map,

$$\mathbf{u} = (u^1, u^2, \ldots u^n) : \left\{ R^{n+1}; \eta_{\mu\nu} \right\} \mapsto \left\{ M^n; g_{jk} \right\} \quad , \tag{0.1}$$

between the Minkowski space R^{n+1} equiped with the metric

$$\eta_{\mu\nu} := \operatorname{diag}\{-1, 1, 1 \ldots, 1\} \quad , \quad \mu\nu = \{0, 1, \ldots n\} \tag{0.2}$$

and some n-dimensional Riemannian manifold M with metric g_{jk}. The map should be the critical point of the following Lagrangian,

$$L(u) := \int_{R^{n+1}} \left\{ g_{jk}(\mathbf{u}) \eta^{\mu\nu} \nabla_\mu u^j \nabla_\nu u^k \right\} d\mathbf{x} \quad . \tag{0.3}$$

Computing the variation of this Lagrangian gives the system of equations

$$\Box u^j + \Gamma^j_{kl}(\mathbf{u}) \left\{ \nabla_\mu u^k \nabla^\mu u^l \right\} = 0 \quad , \tag{0.4}$$

where $\Box := -\nabla_\mu \nabla^\mu$ is the D'Alembertian operator and Γ^j_{kl} are the Christoffel symbols on the manifold M. In the formulation of the problem and in what follows below I will use the summation convention namely summation over repeated indices is imlplied and I raise indices by the formula

$$\nabla^\mu := \eta^{\mu\nu} \nabla_\nu \quad . \tag{0.5}$$

In order to be able to solve (0.4), one has to prescribe intial data on some space-like hypersurface. For example, denoting the first coordinate $x^0 := t$, we can prescribe at $t = 0$ the map $u^j(0, x^1, \ldots, x^n)$ and its time derivatives $u_{,t}(0, x^1, \ldots, x^n)$. Equations (0.4) together with the initial data form the wave map problem. See [12] for a similar but harder problem.

Standard arguments can show that if the initial data are sufficiently smooth, then a smooth solution of (0.4) exists for sufficiently small times. The natural question one can ask is, whether given smooth data the solution of (0.4) exists for all times and if this is not the case to describe the possible loss of smoothness of the map **u**. In general one expects that the behaviour of the solution is determined by an interplay between the dimension n of the underlying Minkowski space and the geometric properties of the target manifold M. This idea is justified to some extend but recent investigations, [2], indicate that if n is large, for example $n \geq 7$, then the geometry of the target manifold is not important and solutions of (0.4) do blow-up in finite time. The most interesting and rich cases are $n = 2, 3$. In general one expects that if the target manifold is negatively curved then solutions of (0.4) are regular, while for positively curved manifold solutions explode in finite time. This is justified to some degree, see [1,5].

There is a distinguished quantity that is crucial for this problem. The best way to introduce it, is through the Energy-Momentum·tensor, namely define the tensor

$$T_{\mu\nu} := g_{jk}u^j_{,\mu}u^k_{,\nu} - \eta_{\mu\nu}\left\{\eta^{\alpha\beta}g_{jk}u^j_{,\alpha}u^k_{,\beta}\right\} \quad , \tag{0.6}$$

then $T_{\mu\nu}$ satisfies,

$$\nabla_\mu T^\mu_\nu = 0 \quad \nu = (0, 1 \ldots n) \quad . \tag{0.7}$$

Now look at the quantity T_{00}. It is a positive quantity and is preserved by the flow i.e. $\int_{R^n} T_{00}d\mathbf{x} = \text{const.}$. This is the total energy of the system. One would like to use this as a first step in order to obtain global solutions. Assume that we can show that a local solution exists for rough intial data that have only finite energy. Then the conservation of energy automatically gives a global solution. This idea can only work for $n = 2$ which is the critical dimension of this problem, see [3].

If we assume that $n = 2$ the problem is critical and there is hope that refinements of existing methods can give definite answers to the problem. Let me rephrase the equations using the language of complex variables. We are seeking a map

$$z(\mathbf{x}) : \left\{R^{2+1}; \eta_{\mu\nu}\right\} \mapsto \left\{D; g(z, \bar{z})dz d\bar{z}\right\} \quad , \tag{0.8}$$

from R^{2+1}, Minkowski space into a subset D of the complex plane C with metric given by the conformal factor $g(z, \bar{z})$. This map is the critical point of the Lagrangian

$$L(z, \bar{z}) := \int_{R^{2+1}} \left\{g(z, \bar{z})\nabla_\mu z \nabla^\mu \bar{z}\right\} d\mathbf{x} \quad , \tag{0.9}$$

and the equation to be solved is

$$\Box z - \left(\log(g(z, \bar{z}))\right)_{,z}\left\{\nabla_\mu z \nabla^\mu z\right\} = 0 \quad . \tag{0.10}$$

There are certain natural conjectures for this problem .

Conjecture 1 : Assume that $g = 4/(1 + |z|^2)^2$, i.e. the target manifold is the unit sphere S^2. For some smooth initial data the solution of (0.10) explodes in finite time in some manner to be described.

Conjecture 2 : Assume that $g = 4/(1 - |z|^2)^2$, i.e. $D := \{|z| < 1\}$ and the target manifold is the hyperbolic plane H^2. The map remains regular provided that the initial data are smooth.

Conjecture 3 : Assume that $n = 3$ and that $M := H^3$ i.e. the hyperbolic three dimensional manifold. The solution of (0.4) is smooth provided that the initial data are smooth.

Conjecture 3 is hopeless at present. However if $M := S^3$ solutions do blow-up in finite time, see [5].

Assuming some symmetry simplifies the problem and certain things are known in these special cases. Assume that $z(t, r)$ where $r := |\mathbf{x}|$ and that the target manifold is H^2 then the solutions of (0.10) are regular provided that the initial data are smooth. This is a remarkable result, proved in [1]. The ouline of the method is illuminating and proceeds as follows. First the authors show that the energy does not "concentrate" inside a characteristic cone. Then they use the fact that the solution can loose regularity only at the central time-line, $r = 0$, in order to estimate derivatives of the map in a weighted L^∞ norm.

There is another symmetry assumption that reduces the problem to a manageable one. Choose coordinates $(t, \mathbf{x}) = (t, re^{i\theta})$ and $z = Re^{i\phi}$. Next make the identification $\phi := k\theta$ where $k =$ integer. This simplification removes the derivatives from the the nonlinear term in (0.10) and the energy controls the range of the map, see [5], [6], [11].

If one wishes to avoid the complications arising from the geometry of the target manifold then the following conjecture must be true.

Conjecture 4 : Assume that the initial data are smooth and the initial energy is sufficiently small. Then solutions of (0.10) remain regular for all time.

The above conjecture states that small energy implies regularity. An even simpler conjecture is the following.

Conjecture 5 : Assume that there exist a smooth solution of (0.10) in the domain $R^2 \times [0, T)$ (this is guaranteed by the local existence theorem, which is standard). Now asume that $z \in C(R^2 \times [0, T])$. Show that z is in fact smooth in $R^2 \times [0, T]$. Hence again by the local existence theorem the solution is smooth in $R^2 \times [0, T_1)$ for some $T_1 > T$.

The conjecture above states that continuity of the map implies regularity and gives some information about the possible way that a solution might loose smoothness.

For the reason that I explained above there is a considerable amount of work done on proving local existence of solutions with as rough initial data as possible. The best result in this direction is [7]. One can outline this method as follows, observe first that,

$$\Box \left\{ \frac{z^2}{2} \right\} = z\Box z - \{\nabla_\mu z \nabla^\mu z\} \quad . \tag{0.11}$$

Then equation (0.10) can be rewritten as

$$\Box z + H(z, \bar{z})\bar{z}\Box z^2 = 0 \quad , \tag{0.12}$$

where H is some function. The construction of the solution is accomplished in the following norm. Consider a function $f(t,x)$ whose Fourier transform I will denote by $\hat{f}(\tau, \xi)$ Now define the norm

$$N_{\delta,s}(f) := \| (1 + ||\tau| - |\xi||)^\delta (1 + |\tau| + |\xi|)^s \hat{f}(\tau, \xi)\|_{L^2(R^{2+1})} \quad . \tag{0.13}$$

This type of norm was introduced in [8], see also [9]. The key ingredient in the construction are two inequalities. For $\epsilon > 0$ arbitrarily small we have

$$N_{\delta,s}(f^2) \leq C N_{\delta,s}^2(f) \quad ; \quad N_{\delta,0}(fg) \leq C N_{\delta,0}(f) N_{\delta,s}(g) \tag{0.14}$$

$$\delta = \frac{1}{2} - \epsilon \quad , \quad s = 1 + 2\epsilon \tag{0.15}$$

See [3], [4]. These inequalities are not optimal, one of the norms on the right hand side should be replaced by L^∞.

References :

1. Christodoulou, D. Tahvildar-Zadeh, A. *On the regularity of spherically symmetric wave maps.* CPAM **XLVI** (1993) , 1041-1091.

2. Cazenave, T. Shatah, J. Tahvildar-Zadeh, A. *Harmonic Maps of the Hyperbolic Space and Development of Singularities in Wave Maps and Yang-Mills Fields.* (Preprint) (1997).

3. Grillakis, G.M. *Apriori Estimates and Regularity of Nonlinear Waves.* Proceedings of ICM Zürich, Switzerland 1995 (1187–1194).

4. Klainerman, S. Machedon, M. *On the regularity properties of a model problem related to wave maps.* Duke Math. Journ. **87** no. 3 553-589, (1997).

5. Shatah, J. Tahvildar-Zadeh, A. *On the Cauchy problem for equivariant wave maps.* CPAM **47** pp. 719-754, (1993).

6. Shatah, J. Tahvildar-Zadeh, A. *Regularity of harmonic maps from the Minkowski space into rotationally symmetric manifolds.* CPAM **45** (1992), 947-971.

7. Tataru, D. *Local and Global solutions for Wave Maps.* Preprint.

8. Beals, M. *Self-spreading and strength of singularities for solutions to semilinear wave equations.* Ann. of Math. **118** 1983 187-214.

9. Bourgain J. *Fourier transform restriction phenomena for certain lattice subsets and applications to nonlinear evolution equations.* GAFA **3** (1993), 209-262.

10. Grillakis, M.G. *Energy Estimates and the Wave Map Problem.* To appear in Comm. in PDE.

11. Grillakis, M.G. *Classical solutions for the equivariant wave map.* (preprint).

12. Weinstein, G. **On Rotating Black Holes in Equilibrium in General Relativity.** CPAM **43** , 903-948.

GEOMETRY OF NODAL SETS AND MULTIPLICITY
OF EIGENVALUES

Nikolai Nadirashvili
University of Chicago

This note is concerned with several problems on the geometry of nodal sets of harmonic functions and related problems on multiplicities of eigenvalues of the Laplace operator on surfaces. Partly these questions are based on the problems raised by S.T. Yau in [16] and [17] (see also [15]). Let us denote by $H_k(*)$ the k-dimensional Hausdorff measure. Let $B_r \subset \mathbb{R}^n$ be the ball $|x| < r$. Let u be a positive harmonic function defined in B_1. Denote by $n_u = u^{(-1)}(0)$ the nodal set of function u. In dimensions greater than 2 little is known about the geometry of n_u. Does the bound

$$(1) \qquad H_{n-1}(n_u) \leq K$$

imply any further "regularity" of the set n_u? For a precise setting of the last question let us denote by N_K the set of all nodal sets of harmonic functions satisfying (1).

Is N_K closed in the Hausdorff topology?

We can pose the last question in a different form:

Conjecture 1. *There are constants C and N depending only on n and K such that if u is a harmonic function in the ball $B_1 \subset \mathbb{R}^n$ and $H_{n-1}(n_u) \leq K$ then*

$$\max_{B_{1/2}} |u| \leq C \sum_{|\alpha|=0}^{N} |D^\alpha u(0)| \ .$$

In dimension 2 the last inequality holds with the constants C and N depending only on the number of nodal domains of the function u [11].

As a weak form of the Conjecture 1 we formulate the next one.

Conjecture 2. *There exists a universal constant $\delta > 0$ such that if u is a harmonic function in $B_1 \subset \mathbb{R}^3$ and $u(0) = 0$ then*

$$H_2(n_u) > \delta.$$

One can also ask if the conclusion of Conjecture 2 holds for solutions of second order elliptic equations with smooth or measurable coefficients in dimensions ≥ 3. In dimension 2, Conjecture 2 is obviously true.

Conjecture 2 implies the following one

Conjecture 3. *Let $G \in \mathbb{R}^3$ be a bounded domain with smooth boundary ∂G. Let u be a Dirichlet eigenfunction:*

$$\Delta u = \lambda u \quad on \quad G,$$

$$u = 0 \quad on \quad \partial G.$$

Then

$$H_2(n_u) > C\lambda^{1/2}$$

Donnelly and Fefferman proved the last inequality for domains bounded by real analytic surfaces, [5].

Let $u \in C^\infty(\overline{B_1})$, be a non-zero solution of a second order elliptic equation with smooth coefficients. Let us denote by E the set of second order zeros of function u. The metric properties of E are of the special interest. Let M be a compact subdomain of B_1. Then $H_{n-2}(E \cap M) < \infty$, [6], [7]. But the structure of the set $E \cap \partial B_1$ is far being understood. The following conjecture is a longstanding problem

Conjecture 4. *Let u be a harmonic function in $B_1 \subset \mathbb{R}^3, u \in C^\infty(\overline{B_1})$,*

$$F = \left\{ x \in \partial B_1 : u(x) = \nabla u(x) = 0 \right\}$$

Then $H_2(F) = 0$.

A C^1 counterexample to the last conjecture was given by Bourgain and Wolff, [2].

We now pass to a discussion of eigenfunctions. Let M be a two-dimensional compact manifold and g be a smooth Riemannian metric on M. Let $u_1, u_2, \ldots$ be eigenfunctions and $0 = \lambda_0 \leq \lambda_1 \leq \ldots$ be the eigenvalues of the Laplace operator on (M, g). Denote by $\mu_k(M, g)$ the multiplicity of λ_k. Set

$$m_k(M) = \sup_g \mu_k(M, g),$$

where sup is taken over all smooth metrics g on M.

In the important paper [3] Cheng proved that for any k and M, $m_k(M) < \infty$. Let M be a two-dimensional sphere $\mathcal{S}^2$. Then

$$
\begin{aligned}
m_1(\mathcal{S}^2) &= 3, \quad [3], \\
m_2(\mathcal{S}^2) &= 3, \quad [8], \\
2\sqrt{k} - 1 \leq m_k(\mathcal{S}^2) &\leq 2k - 1
\end{aligned}
$$

for $k \geq 2$. The left-hand side inequality is given by a standard sphere.

Question. What is the right rate of growth of $m_k(\mathcal{S}^2)$ when $k \to \infty$?

Let M_γ be a surface of genus γ. Then for small γ the number $m_1(M_\gamma)$ is known explicitly, [1], [3], [4], [10], [14]. But for big γ we know only the inequalities

$$c\sqrt{\gamma} < m_1(M_\gamma) < C\gamma,$$

There is again an interesting question: what is the right rate of growth of $m_1(M_\gamma)$ when $\gamma \to \infty$?

Inverse spectral problems are discussed in the conjectures of [16]. (See the appendix, below.) Since the time of that publication, there has been much progress on the problem of describing metrics on the torus whose Laplacian has the same spectrum as the square torus by Zelditch [18], [19]. But the problem remains open.

Estimating L_p norms of eigenfunctions u_k is an important problem. (See [13] and the discussion of the problem in [12].) The case of L_∞ is mentioned in one of the problems in the appendix. Zygmund [20] has shown that on the flat torus $\mathbb{R}^2/\mathbb{Z}^2$ the following inequality holds:

$$||u_k||_4 \leq 5^{1/4}||u_k||_2$$

Question. Can one prove on a surface of negative curvature

$$||u_k||_4 \leq C\lambda_k^\epsilon ||u_k||_2,$$

for any $\epsilon > 0$?

Appendix to
GEOMETRY OF NODAL SETS AND MULTIPLICITY
OF EIGENVALUES [1]

S.-T. Yau
Department of Mathematics
Harvard University

The spectrum for the Laplacian on the two-dimensional sphere is very rigid. There is a good chance that it may determine the metric completely. Recently S. Y. Cheng proved a beautful theorem that if the multiplicity of the spectrum of a metric is the same as the multiplicity of the spectrum of the sphere, then all the geodesics of the metric are closed. In other words, the surface must be a Zoll surface. Can the spectrum of a Zoll surface have the same multiplicity as the spectrum of S^2? Can one say something about the multiplicity of surfaces of higher genus? For example, if the multiplicity of the spectrum of a metric on the torus is the same as the one on the square torus, what can we say about the metric?

The study of the geometry of the eigenfunctions of the Laplacian or the Schrödinger operator is very interesting. The size of the level sets and the critical sets of the eigenfunctions needs to be estimated. For the former quantity, there are excellent works due to Donnelly-Fefferman [5] , Hardt-Simon and Dong. For the latter quantity, essentially nothing is known. Another very important related question is the estimate of the maximum norm of the eigenfunctions when we normalize the L^2 norm to be one. For what types of manifolds can the maximum norm be estimated independent of the size of the eigenvalues?

REFERENCES

[1] G. Besson, Sur la multiplicité de la première valuer propre des surfaces Riemanniennes, Ann. Inst. Fourier 30 (1980), 109-128.
[2] J.Bourgain, T.Wolff, A remark on gradients of harmonic functions in dimension 3, Colloq. math. 60/61 (1990), 253-260.
[3] S.Y. Cheng, Eigenfunctions and nodal sets, Comment. Math.Helv. 51 (1976), 43-55.
[4] Y. Colin de Verdière, Sur la multiplicité de la première valeur propre non nulle du Laplacien, Comment. Math. Helv. 61 (1986), 254-273.
[5] H.Donnelly, Ch. Fefferman Nodal sets of eigenfunctions: Riemannian manifolds with boundary, in Analysis et cetera, Academic Press 1990, 251-262.

---

[1] The following are problems 41 and 43 reprinted from [16].

[6]   R. Hardt,M. Hoffman-Ostenhof, T. Hoffman-Ostenhof, N. Nadirashvili, Critical sets of solutions to elliptic equations, J. Diff. Geom. (1999).

[7]   M. Hoffman-Ostenhof, T. Hoffman-Ostenhof, N. Nadirashvili, Critical sets of
smooth solutions of elliptic equations in dimension 3, Indiana Univ. Math. J.
45 (1996), 15-37

[8]   M. Hoffman-Ostenhof, T. Hoffman-Ostenhof, N. Nadirashvili, On the multiplicity of eigenvalues of the Laplacian on surfaces, Ann. Global Analysis and
Geometry 17 (1998), 43-48.

[9]   T. Hoffman-Ostenhof, P.Michor, N.Nadirashvili, Bounds on the multiplicity of
eigenvalues of fixed membrane, Preprint, 1998.

[10] N. Nadirashvili, Multiple eigenvalues of Laplace operator, Math. USSR Sbornik
61 (1988), 225-238.

[11] N.Nadirashvili, Harmonic functions with bounded number of nodal domains,
Appl. Analysis, (1998).

[12] P.Sarnak, Spectra and eigenfunctions of Laplacians, Centre de Recherches
math. CRM Proc. and Lect. Notes 12 (1997), 261-276

[13] A.Seeger, C.D.Sogge, Bounds for eigenfunctions of differential operators, Indiana Univ. Math. J. 38 (1989), 669-682.

[14] B. Sevennec, Majoration topologique de la multiplicité du spectre des surfaces,
Séminaire de Théorie spectrale et géométrie, N12, Anée 1993-1994, Grenoble.

[15] R.Schoen, S.T.Yau, Lectures on Differential geometry, International Press,
1994

[16] S.T. Yau, Open problems in geometry, Proc. Symp. Pure Math. 54, 1 (1993),
1-28. Reprinted in Chern: A Great Geometer of the Twentieth Century, IP,
1992, 275-319.

[17] S.T. Yau, Problem section, Ann. Math. Study, 102 (1982), 669-706.

[18] S.Zelditch, Maximally degenerate Laplacians, Ann. l'Institut Fourier 46 (1996),
547-587.

[19] S.Zelditch,Fine structure of Zoll spectra, J. Fun. Anal. 143 (1997), 415-460.

[20] A.Zygmund, On Fourier coefficients and transforms of functions of two variables, Studia Math. 50 (1974), 189-201.

Questions related to the dimension of Kakeya sets
Thomas Wolff

The problems we will mention are all fairly well-known and none of them originate with the author. Roughly speaking they come from an area where harmonic analysis interacts with certain aspects of geometric measure theory or combinatorial geometry. They are discussed in more detail in several expository articles in the literature, most recently in [9] which contains proofs or references for most of the statements made below. Relevant general references are [3] and [8].

Perhaps the most basic is the following form of the Kakeya maximal conjecture: define a maximal operator from functions on $\mathbb{R}^n$ to functions on the sphere depending on a parameter $\delta > 0$ in the folowing way: if $e \in S^{n-1}$, $a \in \mathbb{R}^n$ then let

$$T_e^{\delta}(a) = \{x \in \mathbb{R}^n : |(x-a) \cdot e| \le \frac{1}{2}, |(x-a)^{\perp}| \le \delta\}$$

where $x^{\perp} = x - (x \cdot e)e$. Thus $T_e^{\delta}(a)$ is essentially the δ-neighborhood of the unit line segment in the e direction centered at a. If $f : \mathbb{R}^n \to \mathbb{R}$ then (see Bourgain [1]) we define its Kakeya maximal function $f_{\delta}^* : S^{n-1} \to \mathbb{R}$ via

$$f_{\delta}^*(e) = \sup_{a \in \mathbb{R}^n} \frac{1}{|T_e^{\delta}(a)|} \int_{T_e^{\delta}(a)} |f|$$

It follows from existence of Kakeya sets (compact sets with measure zero containing a unit line segment in each direction) that it is impossible to make an $L^p \to L^p$ estimate for this maximal function with a bound

237

independent of δ, and the question is whether one can obtain less than power dependence on δ:

Question 1: Is there an estimate

$$\forall \epsilon > 0 \exists C_\epsilon : \|f_\delta^*\|_{L^p(S^{n-1})} \leq C_\epsilon \delta^{-\epsilon} \|f\|_{L^p(\mathbb{R}^n)} \forall f, \text{ where } p = n \qquad (1)$$

A somewhat weaker but morally equivalent version of this question asks whether a Kakeya set in $\mathbb{R}^n$ must have Hausdorff dimension n. This would follow ([1], Lemma 2.15) from an affirmative answer to question 1.

Many techniques are known [8] for proving maximal inequalities like (1) - they typically involve either combinatorial- geometric techniques, Fourier analysis techniques based on the Plancherel theorem or some combination thereof. It is well known that no estimate like (1) can hold on L^p with $p < n$ (consider f = characteristic function of a ball of radius δ). This makes (1) seem hard to approach via Fourier analysis techniques with the exception of the case $n = 2$ where (1) is known and can be proved in a page using either Fourier analysis [1] or the geometrical argument of A. Cordoba. In higher dimensions it is not hard to show using geometrical or semi geometrical arguments that Kakeya sets have dimension $\geq \frac{n+1}{2}$ and there has been a certain amount of progress past this point, beginning with [1]. The current best result implies Kakeya sets have dimension $\geq \frac{n+2}{2}$, see [9] and the references there. These results are still based on fairly straightforward ad hoc geometrical arguments and it appears hard to go further with such an approach. Further remarks on some of the issues involved are in [9] and at the end of [10]. Schlag's paper [6] is also relevant here.

Let us point out that there are a large number of unsolved problems in several areas whose solution seems dependent on being able to answer question 1.

Many of them are questions in Fourier analysis which are connected to (1) via the "disc multiplier" argument of C. Fefferman [4] and related considerations. The simplest problem to state and the easiest to connect to (1) is the well known

Restriction conjecture of Stein: let μ be a measure on the unit sphere in $\mathbb{R}^n$ which has a bounded density with respect to surface measure and

238

let $\hat{\mu}$ be its Fourier transform. Then is there an estimate $\hat{\mu} \in L^p \; \forall p > \frac{2n}{n-1}$?

It is known (e.g. [2]) that this estimate would imply (1) by a variant on the argument of [4].

On the other hand it is unclear whether (1) would imply the restriction conjecture; a proof of such a conditional result would presumably be a very significant advance. It is known that (1) (or improved partial results on (1)) leads to improved partial results on e.g. the restriction problem. This is due to Bourgain [1] and [2] and there has been some further recent related work by Moyua-Vargas-Vega and by Tao.

In addition, there are a number of questions of a geometrical nature which are closely related to (1). A classical problem motivated by Besicovitch's construction of Kakeya sets with measure zero is the following:

Problem of $(n, 2)$ sets: does there exist a set in $\mathbb{R}^n$ with measure zero which contains a translate of every 2-plane?

Such a set would be called a Besicovitch $(n, 2)$-set. One can also formulate this question more quantitatively in terms of the maximal 2-plane transform. Further, one can ask the analogous question for k-planes, $k \geq 3$ and there is no value of k for which a negative answer is known for all n. For example, for every $n \geq 5$ it is unknown whether $(n, 2)$-sets exist. An argument in [1] shows that if the answer to question 1 is affirmative then there is no $(n, 2)$ set for any n.

Next we mention a question in $\mathbb{R}^2$; so far as the author knows there is no implication between this question and (1), but it is natural to expect that they are related.

Suppose that $\alpha \in (0, 1]$ and that E is a compact set in $\mathbb{R}^2$ with the following property: for every direction e there is a line segment in the e-direction which intersects E in Hausdorff dimension $\geq \alpha$. Then what is the optimal lower bound for the Hausdorff dimension of E?

This question arose from considerations in [5] (the author is grateful to Yuval Peres for this reference). It is of course possible to prove various partial results: so far as the author is aware, what is known is that $\dim E \geq \max(2\alpha, \frac{1}{2} + \alpha)$ and that $\dim E$ can be as small as $\frac{1}{2} + \frac{3}{2}\alpha$.

239

The latter fact is essentially an example of Erdos (of an elementary number theoretic nature) which also shows sharpness of some related combinatorial results.

Finally let us mention a question in $\mathbb{R}^2$ of a different nature, somewhat analogous to the question of whether (1) would imply the restriction conjecture. This question originates in [7]. Namely, consider the initial value problem for the wave equation in n space variables, where in principle $n \geq 2$ is arbitrary: $\Box u = 0, u(\cdot, 0) = f, \frac{\partial u}{\partial t}(\cdot, 0) = 0$.

Then is there an estimate

$$\forall \epsilon > 0 \exists C_\epsilon : \|u\|_{L^p(\mathbb{R}^n \times [1,2])} \leq C_\epsilon \|f\|_{p,\epsilon} \tag{2}$$

when $p \in [2, \frac{2n}{n-1}]$? Here $\|\cdot\|_{p,\epsilon}$ is the inhomogeneous L^p Sobolev norm with ϵ derivatives.

This can be regarded as a souped up version of the restriction conjecture and is known to imply the restriction conjecture, hence (1), and also the well-known Bochner-Riesz conjecture. Hence it is presumably quite difficult in three or more dimensions. On the other hand, when $n = 2$ the restriction and Bochner-Riesz conjectures are known to hold [8] but (2) is nevertheless open. We note that (2) with $n = 2$ encodes some geometric-combinatorial information (beyond the easy two dimensional case of (1)). In particular it would imply the following fact: let E be a set in $\mathbb{R}^2$ which contains a circle of every radius. Then E must have Hausdorff dimension 2. This fact can be given an independent proof, which is nontrivial but is based on known combinatorial arguments. But the latter arguments do not readily give a proof of the $n = 2$ case of (2), which could be a good testing ground for new techniques in this area.

References

[1] J. Bourgain, *Besicovitch type maximal operators and applications to Fourier analysis*, Geometric and Functional Analysis 1 (1991), 147-187.

[2] J. Bourgain, *Some new estimates for oscillatory integrals,* in *Essays on Fourier analysis in honor of Elias M. Stein,* ed. C. Fefferman, R. Fefferman, S. Wainger, Princeton University Press, 1994.

[3] K. J. Falconer, *The geometry of fractal sets,* Cambridge University Press, 1985.

[4] C. Fefferman, *The multiplier problem for the ball,* Ann. Math. 94(1971), 330-336.

[5] H. Furstenburg, *Intersections of Cantor sets and transversality of semigroups,* in *Problems in Analysis: a symposium in honor of Salomon Bochner,* ed. R. Gunning, Princeton University Press, 1970.

[6] W. Schlag, *A geometric inequality with applications to the Kakeya problem in three dimensions,* Geometric and Functional Analysis, to appear.

[7] C. Sogge, *Propagation of singularities and maximal functions in the plane,* Inv. Math. 104 (1991), 349-376.

[8] E. M. Stein, *Harmonic Analysis,* Princeton University Press, 1993.

[9] T. Wolff, *Recent work connected with the Kakeya problem,* Anniversary Proceedings, Princeton 1996, to appear.

[10] T. Wolff, *A mixed norm estimate for the x-ray transform,* to appear.

Thomas Wolff
253-37 Caltech
Pasadena, Ca 91125, USA

Open Problems in Partial Differential
Equations Arising from Fluid Dynamics

Zhouping Xin[1]

Courant Institute

New York University

Mathematical theory of fluid dynamics gives rise a galaxy of well-known nonlinear partial differential equations such as the celebrated Euler equations and Navier Stokes equations for the compressible and incompressible fluids ([1, 2]), nonlinear Boltzmann equations for rarefied gases ([3]) and their variations by taking into account of various additional physical effects. The fundamental importance in various applications in many different branches of sciences and the richness in the mathematical theory of these nonlinear systems have made them some of focuses of extensive researches in the last few decades by mathematicians and engineers, and tremendous progress has been made in solving these equations both theoretically and numerically, and in understand-

---

[1]Address: Courant Institute, NYU, 251 Mercer Street, N.Y. N.Y. 10012 email: xinz@cims.nyu.edu

ing the behavior of solutions to such nonlinear systems. Despite the substantial success and great effort by many mathematicians in the development of the mathematical theory for these nonlinear partial differential equations ([1, 3, & 4]), there remain many fundamental open problems which challenge the field now and in the future. In the following, I will list some of the open problems which, in my opinion, should be of importance in understanding these nonlinear partial differential equations.

§1. Compressible Euler Equations and the Mathematical Theory of Shock Waves

The system of Euler equations for compressible fluid ([1]) provides one of the most important examples of quasi-linear hyperbole conservation laws which is a notoriously difficult subject due to the richness and complexity caused by the fact that the wave speeds depend on the wave themselves. The past progress has been concentrated on one-dimensional theory with very few exceptions (see [4, 5, 6] and the reference therein). There remain many important problems to be solved. We start with one-dimensional problems.

1. Global (in time) well-posedness of entropy weak solutions to the Cauchy problems of the one-dimensional Euler equations for polytropic gas dynamics in the space of functions with bounded total variations (which is *BV* space). Even though the global existence of solutions in *BV* space has been obtained by the Glimm's scheme ([7, 4]), and their uniqueness and continuous dependence in the class of the viscosity solutions ([8]) has been shown ([8])

for initial data with small total variations, all the relevant questions remain open for solutions with large data in BV space. The significance of this problem lies in the facts that the BV space is natural space of basic waves such as shock waves, centered rarefaction waves, and contact discontinuities to develop and interact, and the physical relevance of strong linear and nonlinear waves ([1]). In fact, this problem has not been solved even for isentropic gases with general data in BV without the additional assumption of initial small oscillations ([13]), although in this case large amplitude weak solutions exist in the space of bounded measurable functions thanks to the theory of compensated compactness ([10, 11, 12]). However, little information is available on the structures of such solutions. Based on the insights provided by the studies of weakly nonlinear geometric optics approximations, there are recent interesting results which show finite time blow-up in BV norm for solutions to some systems of nonlinear hyperbolic conservation laws with initial data in BV space of small oscillation ([13, 14, 15]). However, such systems have different structures from those of Euler equations for polytropic fluids. Thus it is extremely interesting to know whether a weak solutions in BV space can blow-up in finite time for gas dynamics equations with large data in BV.

2. Uniqueness of the vanishing viscosity limits. Consider a weak solution to the one-dimensional Euler equations for polytropic fluids, which is a limit (in appropriate topology) of solutions of the corresponding Navier-Stokes equations as the dissipation parameters (viscosity and heat conduction coefficients) tend to zero.

Show that such a solution is the unique weak solution to the Cauchy problem for th Euler equations under appropriate entropy conditions. A less ambitious problem is to prove that such a solution is unique among the class of weak solutions which can be obtained as limit of solutions to the corresponding Navier-Stokes equations (or even the system of Euler equations with artificial viscosities) as dissipation goes to zero. This has been established only for very special cases ([16, 17]). One possible approach is to compare the vanishing viscosity limit solutions with the viscosity solutions defined in ([8]). This requires some deeper understanding of the asymptotic local behavior of solutions to the Navier-Stokes equations as dissipation becomes small. The recently developed theory of viscous shock waves ([18–24]) should yield considerable insight and provide some of the important analytical tools to settle this problem. A related problem is to show the uniqueness of weak solutions to Euler system for isentropic fluids for which the vanishing viscosity solutions have been obtained by the theory of the compensated compactness ([10, 12]).

3. Spatial periodic solutions. In their fundamental work [25], Glimm and Lax obtained the existence and large time asymptotic behavior of spatial periodic solutions to the isentropic Euler equations. Their results display the amazing regularization effects of the genuined nonlinearity which damps out any initial oscillations, which lays the rigorous foundation for the later-on homogenization theory on the propagation of high frequency waves for this system. Despite its extreme importance and much effort in the past two decades, the theory of global (in time) existence and large time

asymptotic behavior of spatial periodic solutions to the full Euler equation for polytropic fluids remains completely open. Main difficulties include the lack of Riemann invariant coordinates and the resonant wave interactions through the entropy wave field which is linearly degenerate. Some of related but simpler questions are: Is there example of spatial periodic weak solution for the full Euler equation with piecewise constant periodic initial data which ceases to exist in finite time? Are there examples of special spatial periodic solutions which decay to their average values over each period with different rate of decay in time?

4. Local structures of solutions in BV space.

Let $V(x,t) \in BV$ be a weak solution to the full Euler equations of polytropic fluids satisfying the physical entropy condition. What are the additional local structures of $V(x,t)$ in the physical space beyong those of a BV functions ([26])? Can $V(x,t)$ behave locally as an appropriate perturbation of a Riemann solutions as Glimm solutions do? In particular, can $V(x,t)$ be an viscosity solution in the sense of [8]? Some interesting results have been obtained by Dafermos in this directions for the isentropic Euler equations based on the generalized characteristic methods and entropy estimates [27].

We next turn to the multi-dimensional theory of shock waves which is an extremely important but difficult subject. There has been very little progress for the multi-dimensional shock wave theory for the compressible Euler system with a few exceptions (see [1] and [5] and the references therein). Some completely new ideas and techniques are needed to lay a foundation for the multi-dimensional shock wave theory. Some of the following open problems serve as steps to achieve better understanding of this great task. We also refer the readers to [5, 28] for some related perspective and open problems in this direction.

5. Structures of singularities and formation of shocks. The inevitable formations of shock waves in finite-time of solutions with generic smooth initial data is one of the most striking features of nonlinear hyperbolic conservation laws. This has been pretty well understood for one-dimensional Euler equations for polytropic fluids ([6, 29, 30]). In higher space dimensions, it is known that smooth solutions will blow-up in finite time for compressible Euler equations with generic smooth initial data ([31, 32]). However, the nature of singularities in the solutions is not clear except for special cases for which the singularities are shown to be shocks (see [33] and the reference therein). It is important to characterize the structures of singularities forming from smooth solutions in terms of initial data. In particular, it is of great interest to determine when shocks form first before the development of shell singularities.

6. The well-posedness of weak solutions. The well-posedness of weak solutions to the compressible Euler equations in higher space dimensions remains to be one of the most challenging open problems for mathematical theory of compressible fluid dynamics. It is not even clear that in which functional space the problem should be posed in contrast to the one-dimensional theory for which the spaces BV is natural (see [34]). Even the short time existence of piecewise smooth solutions are not proved except for the special cases that the initial data are suitable perturba-

tions of shock fronts or planary rarefaction waves (see [5] and the reference therein). Thus there are many basic questions to be answered. What are the geometric structures of generic discontinuities in a weak solution? How do the basic linear and nonlinear waves evolve and interact? What are the appropriate entropy conditions? Are the planary shock waves (satisfying the entropy conditions in [5]) and centered rarefactions waves asymptotically stable under non-planary perturbations? What is the appropriate space in which one can obtain either existence (even for short time) or uniqueness of general physically relevant weak solutions? When is the solution operator compact? Due to the complexity and the lack of understanding of the general problem, it might be very helpful to study some of the special physically relevant wave patterns for which a lot experimental data, numerical simulations, and formal asymptotics results are available. Some of the well-known important examples include: shock wave reflection phenomena ([35]), transonic flows and standing shocks in a nozzle [1], self similar flows, and other flows with various symmetry properties.

§2. Compressible Navier-Stokes Equation and Viscous Shock Wave Theory

Homogeneous Newtonian fluids are described by the compressible Navier-Stokes system when the viscosity and heat conduction are taken into account. This is a hyperbolic-parabolic composite type quasi-linear system with dissipative structure. Substantial progress has been achieved in the past several decades in understanding the solutions to the compressible Navier-Stokes equations. In particular, the well-posedness and large time asymptotic behavior towards constant states of small amplitudes smooth (or even *BV*) solutions in Sobolev spaces to the compressible Navier-Stokes equations for fixed viscosity and heat-conductivity have been well-established (see [36], [37]), the one-dimensional theory of nonlinear asymptotic stability toward linear and nonlinear waves has also been completed (see [18–24]), and even the existence of weak solution of large amplitudes to the 2D or 3D Navier-Stokes system for some isentropic fluids has been announced in [38]. However, there remain many more open problems which are of significance both mathematically and physically, such as the asymptotic equivalence in the limit of small dissipation between the compressible Navier-Stokes system and the Euler equations in the presence of discontinuities and boundaries; existence (or non-existence) of large amplitude general solutions; and large time asymptotic behavior toward basic waves, etc.

7. Global (in time) well-posedness of the Cauchy problem or initial-boundary value of problem for the compressible Navier-Stokes equations with large data. Consider the Navier Stokes equations for viscous polytropic flows with fixed positive viscosity and non-negative heat conductivity without exterior forcing. Is there formation of singularities in finite time (or even in infinite time) for the solutions with large smooth data ([39])? Can the vacuum state be formed in finite time even the data are given away from the vacuum? If there exist singularities forming from smooth solutions, can one characterize the structures of the singularity as in the case for harmonic maps? What are the appropriate spaces to study the weak solutions after formations of singularities? What

are the large time asymptotic behavior of the weak solutions?

8. Large time asymptotic states for the multi-dimensional compressible Navier-Stokes equations. The theory of large time asymptotic behavior of small solutions to the one-dimensional Navier-Stokes systems for viscous polytropic fluids is almost completely understood now in the sense that viscous shock waves ad rarefaction waves are nonlinearly stable and contact discontinuities are metastable with detailed pointwise asymptotic ansatz constructed and justified (see [17–24]). However, in higher space dimensions, only constant states are shown to be stable ([36]) except some scalar models ([40, 41, 42]). It is of great interest to know whether basic nonlinear wave patters such as planar viscous shock profiles and viscous rarefaction waves, and contact waves are nonlinearly stable under generic multi-dimensional perturbations. Can one determine the leading order (or even higher order) asymptotic ansatz of a solution to the compressible Navier-Stokes equations when its initial data are small perturbations of these basic wave patterns? These questions are not only important in the stability theory itself, but also should share light on the asymptotic equivalence between the compressible Navier-Stokes equations and Euler system in the limit of small dissipation which will be discussed in the next problem.

9. Asymptotic equivalence between Navier-Stokes system and Euler equations for compressible fluid in the limit of small dissipation. In the limit of small viscosity and heat conductivity, the solutions to the compressible Navier-Stokes equations display turbulent behavior whose understanding is still one of major challenging prob-

lems in fluid dynamics. This is particularly so if either the corresponding Euler flows contain discontinuities or there appear solid boundaries. Some of the main difficulties are due to the dramatic change of gradients of the flows across the discontinuities and near boundaries where the small scale effects are strong. In the case that the underlying Euler flows are piecewise smooth with finitely many entropy satisfying shock discontinuities, it is exported that solutions to the Navier-Stokes equation will converge uniformly to the corresponding solutions to the Euler equations except around thin layers near shock discontinuities in the zero dissipation limit, and furthermore, the precise asymptotic ansatz of the solutions to the Navier-Stokes system for small but nonzero dissipation can be determined by multi-scale matched asymptotic analysis. This has been proved only for one-dimensional problems ([23]). The case that the solutions to the Euler equation are piecewise smooth containing contact discontinuous surfaces is more subtle due to the instability of vortex sheets ([2]). The behavior of solutions around a contact discontinuity surface for multi-dimensional Navier-Stokes equations for small dissipation is open even formally. In the presence of physical boundaries, the well-known boundary layer phenomena occur due to discrepancy of the non-slip boundary conditions for the Navier-Stokes system and the prescribed normal component of the velocity field at the boundaries for the Euler equations ([2, 43]). Formal asymptotic analysis yields the well-known Prandtl's laminar boundary layer theory whose validity and rigorous justification are completely open in most physically relevant flows except for linearized Navier-Stokes equation ([44]). This is so even for short time before the Euler flows develop any singularities. There have been extensive studies on the steady Prandtl's boundary layer equations (see [45, 46]). Yet, very little is known for dynamical boundary layers. Even the short time well-posedness in Sobolev spaces of the unsteady Prandtl's boundary layer equations has not been achieved (but see [47, 48]).

§3. Incompressible Euler and Navier-Stokes equations

In the limit of small Mach numbers, one derives from the compressible Euler and Navier-Stokes systems the incompressible Euler and Navier-Stokes equations respectively ([2, 49]), which govern various flows such as waters in the oceans and air in the atmosphere. The mathematical theory for the incompressible fluid is more mature than that for the compressible fluid (see [50, 51, 52]). In particular, a rather complete well-posed theory exists in two space dimensions for both Navier-Stokes and Euler equations except the important non-smooth flow—vortex sheets; and under minimal assumptions on the data and exterior forcing, the Leray-Hopf weak solutions to the 3-dimensional Navier-Stokes exist globally in time ([50]), and the partial regularity of the weak solutions for 3-D Navier-Stokes equation ([53]), large time decay of the solutions ([54]), and existence of large strong solutions under the assumption of certain spatial symmetries (see [55] and reference therein), have been established. Yet many of the central problems in incompressible fluid mechanics are still challenging the field. We will mention only a few well-known problems here.

10. Well-posedness of the Cauchy problem for the incompressible Euler equations. Do smooth solutions to the 3-dimensional incompressible Euler equations break down in finite time? A positive answer to this question has a profound physical implication that the singularities signify the onset of turbulence, and their structures should share light on the internal cascade in turbulent flows at high Reynolds numbers. This problem has been studied extensively by both analytical and numerical methods in the past several decades which strongly suggest the possible breakdown in finite time of the smooth solution (see [52], and the reference therein). Yet the formation of singularities in finite time has not been rigorously confirmed so far. The lack of understanding of the structures of the possible singularities also prevents one from obtaining the well-posedness of weak solutions to the 3-D Euler equations even though the existence of measure-valued solutions has been proved [57]. The principal mechanism, responsible for the possible formations of singularities, is the stretching and accumulation of the vorticity (see [52]). In the cases that there is no stretching of vorticity, such as 2-dimensional flows and axisymmetric flows without swirls, smooth global (in time) solutions to the Euler equations are obtained [52, 58].

11. Vortex sheets problem and concentration of energy for 2-dimensional Euler equations. Although 2-dimensional smooth inviscid incompressible flows are very well understood to some extent, yet many important questions involving 2-dimensional singular flows remain to be answered. This is particularly so in the case of the vortex sheets problem, in which the vorticity is a finite Radon

measure concentrated or a curve which is initially smooth, and the corresponding velocity field has locally finite energy. This gives a classical example of ill-posed initial value problem in the sense of Hadamard. At later time, a singularity in the sheet may develop and the nature of solutions past the singularity formation is unknown [59]. The existence of classical weak solutions to the 2-D Euler equations with vortex sheets initial data has been achieved by the concentration-cancellations technique ([56, 57]) under the additional assumptions that the initial vorticity has one sign [50]. It is of fundamental importance both physically and mathematically to know whether such a concentration-cancellation occurs to yield the existence of classical weak solution (in contrast to measure-valued solutions) to 2-D Euler equations with general vortex sheets initial data. Another related problem is to understand the structures of the approximate solutions to the vortex sheets problem generated by either viscous approximation or vortex methods [56], which is of considerable significance both physically and numerically. It has been shown ([61, 62]) that concentration-cancellation always occurs for such approximate solutions to vortex sheets problem provided that the initial vorticity has one sign. Does the same phenomena occur for general vortex sheets initial data? Does energy-concentration occur at all dynamically for such approximate solutions in the sense that the energy-concentration occurs later on in time even though the initial data for the approximate solutions converges strongly to that of the initial vortex sheets data in energy norms. The later question is open even in the case that the initial vorticity has one sign.. Finally, the uniqueness of weak solutions to the 2-D Euler equations are wildly open except the case where initial vorticity are bounded in the super-norm [52].

12. Regularity and asymptotic behavior of solutions to the incompressible Navier-Stokes equations. Although there is no evidence either numerically or physically to support the breaking down of smooth solutions to the 3-D incompressible Navier-Stokes equations, yet the global (in time) existence of smooth to the 3-D incompressible Navier-Stokes system is one of the most challenging mathematical problems to be solved. This is related to the regularity of Leray-Hopf weak solution [50], for which only partial regularity, results are available (see [53, 64, 65]). There are recent interesting results which shows self-similar blow-up (Leray's self-similar solutions) are impossible [66]. These problems are open even for general viscous axisymmetric flows. Other important open problems include: asymptotic behavior of solutions to the Navier-Stokes equations either as time approaches infinitely or as Reynold number goes to infinity: do inertial manifolds exist for 2-D or 3-D Navier-Stokes equations [50]? Is Prandtl's boundary layer theory valid for the viscous incompressible fluids in the presence of boundary [43, 46]? What are the asymptotic forms of the solutions to the Navier-Stokes system as the Reynold number goes to infinity in the case of corresponding Euler solutions are singular?

§4. Nonlinear Boltzmann Equation

The general Bolzmann equations of kinetic theory gives a statistical description of a gas of interacting particles. An important property of this equations is its asymptotic equivalence to the Euler and Navier-Stokes systems of compressible fluid dynamics in the limit of mean free path ([3, 67]). It is expected physically that away from initial layers, shock layers, and boundary layers, the solution to the Boltzmann equation should relax to its equilibrium states (local Maxwellian state) in the limit of small mean-free path so that the gases are governed the macroscopic equations such as Euler or Navier-Stokes equations ([3]). This is predicted by the method of normal solutions (or normal region) based on the Hilbert expansion and the Chapman-Enskeg expansions. The rigorous mathematical justification of this fluid-dynamic approximations of Boltzmann solutions poses a challenging open problem in most important cases, in particular, in the case that there are shock discontinuities in the fluid flow and/or in the presence of boundaries. There has been extensive studies on this hydrodynamic limit problem. However, most work deal with smooth flows and without boundaries, or some model linearized Boltzmann equation (see [67, 3, 68, 69, 70]) with exceptions [71–74]. A qualitative theory for boundary layer problem exists for some models for steady Boltzmann equations [75], very little is know for unsteady problems. The hydro-dynamic limit problem in the presence of discontinuities in the fluid flows is completely open for general nonlinear Boltzmann equation except for Broadwell model [72, 76]. Even a formal asymptotic theory for the shock layer and boundary layer problems has not been achieved. One of the main difficulties in analyzing these problems is due to the complexity of the nonlocal collision operator in the Boltzmann equation, which makes it difficult to study the structures of the layer problems associated with the formal matched asymptotic analysis. Even when the formal asymptotic solutions can be constructed such as for the Broadwell model [72], the convergence analysis is still highly nontrivial due to the fact that the fluid-dynamic limits are strongly singular. However, since shock waves are essential for compressible fluids and boundary layer theory describes the interactions of the gas molecules with the molecules of the solid body, to which one can trace the origin of the drag exerted by the gas on the body and the heat transfer between the gas and the solid boundaries, it is of fundamental importance to understand the fluid dynamic approximation to the nonlinear Boltzmann equation in the presence of shocks and boundaries. Other related questions, such as global existence of large amplitude smooth solution to the general nonlinear Boltzmann equation [3], and the regularity of the renormalized weak solutions to Boltzmann equations due to Diperna–Lions ([77]) and its hydro-dynamic limit, are waiting to be attacked too.

References

[1] Courant, R. and Friedrich, K. O., *Supersonic Flow and Shock Waves*, Wiley-Interscience, 1948.

[2] Landau, L. and Lifschitz, E., *Fluid Mechanics*, Addison-Wesley, New York, 1953.

[3] Cercignani, C., *The Boltzmann Equation and Its Applications*, Springer-Verlag, New York, 1988.

[4] Smoller, J., *Shock Waves and Reaction-Diffusion Equations*, Berlin–Heidelberg–New York: Springer Verlag, 1982.

[5] Majda, A., *Compressible Fluid Flow and Systems of Conservation Laws*, Applied Mathematical Sciences, **53**, Springer-Verlag, New York, 1984.

[6] Lax, P. D., *Hyperbolic Systems of Conservation Laws and the Mathematical Theory of Shock Waves*, SIAM Regional Conference Series in Appl. Math., no. 11 (1973), Philadelphia.

[7] Glimm, J., *Solutions in the large for nonlinear hyperbolic systems of equations*, Comm. Pure Appl. Math., vol. **18** (1965), 95–105.

[8] Bressan, A., *The unique limit of the Glimm scheme*, Arch. Rat. Mech. Anal. **130** (1995), 205–230.

[9] Temple, B., *Solutions in the large for some nonlinear hyperbolic conservation laws of gas dynamics*, J. Diff. Eqn., **41** (1981), 96–161.

[10] Diperna, R., *Convergence of the viscosity methods for isentropic gas dynamics*, Comm. Math. Phys. **91** (1983), 1–30.

[11] Ding, X., Chen, G. Q., and Luo, P., *Convergence of the fractional step Lax–Friedrichs and Godunov scheme for the isentropic gas dynamics*, Comm. Math. Phys. **121** (1989), 63–84.

[12] Lions, P., Perthame, B., and Souganidis, P. E., *Existence and stability of entropy solutions for the hyperbolic systems of isentropic gas dynamics in Eulerian and Lagrangian coordinates*, Comm. Pure Appl. Math. **49** (1996), 599–638.

[13] Temple, B. and Young, R., *Solutions to the Euler equations with large data*, Proceedings of the Fifth International Conference on Hyperbolic Problems, Theory, Numerics, Applications, eds. by Glimm, etc. World Scientific, (1996), 258–267.

[14] Metivier, G., *A nonlinear instability for 3 × 3 systems of conservation laws*, Preprint, 1993.

[15] Hunter, J., *Strongly nonlinear hyperbolic waves. Nonlinear Hyperbolic Equations—Theory, Computational Methods, and Applications*, J. Ballman, etc. (eds.) Viewig, 1989, 257–268.

[16] Hoff, D. and Liu, T. P., *The inviscid limit for the Navier-Stokes equations of compressible, isentropic flow with shock data*, Indiana Univ. J., **38** (4) (1989), 861-915.

[17] Xin, Z., *Zero dissipation limit to rarefaction waves for the one-dimensional Navier-Stokes equations for compressible isentropic gas*, Comm. Pure Appl. Math., **86** (1993), 1499–1533.

[18] Goodman, J., *Nonlinear asymptotic stability of viscous shock profiles for conservation laws*, Arch. Rat. Mech. Anal., **95** (1986), 325–344.

[19] Liu, T. P., *Shock waves for compressible Navier-Stokes equations are stable*, Comm. Pure Appl. Math., **39** (1986), 565–594.

[20] Liu, T. P. and Xin, Z., *Nonlinear stability of rarefaction waves for compressible Navier-Stokes equations*, Comm. Math. Phys., **118** (1988), 451–465.

[21] Szepessy, A. and Xin, Z., *Nonlinear stability of viscous shock waves*, Arch. Rat. Mech. Anal., **122** (1993), 53–103.

[22] Liu, T. P. and Xin, Z., *Pointwise decay to contact discontinuities for systems of conservation laws*, Asian J. Math., vol. **1** (1997), 34–84.

[23] Goodman, J. and Xin, Z., *Viscous limits to piecewise smooth solutions of systems of conservation laws*, Arch. Rat. Mech. Anal., **121** (1992), 235–265.

[24] Liu, T. P., Preprint, 1997.

[25] Glimm, J. and Lax, P., *Decay of solutions of systems of nonlinear hyperbolic conservation laws*, AMS Memoirs **101** (1970).

[26] Velpert, A., *The spaces BV and quasilinear equations*, Mat. Sb., **73** (1967), 255–302; or Math. USSR, Sb., **2** (1967), 225–267.

[27] Defermos, C. M., *Admissible wave fans in nonlinear hyperbolic systems*, Arch. Rat. Mech. Anal., **106** (3) (1989), 243–260.

[28] Majda, A., *One perspective on open problems in multidimension conservation laws,*

[29] Liu, T. P., *Development of singularities in the nonlinear waves for quasilinear hyperbolic partial differential equations*, J. Diff. Eqns., **33** (1979), 92–111.

[30] John, F., *Formation of singularities in one-dimensional nonlinear wave propagation*, Comm. Pure Appl. Math., **27** (1974), 337–405.

[31] Sideris, J., *Formation of singularities in three-dimensional compressible fluids*, Comm. Math. Phys., **101** (1985), 475–487.

[32] Makino, T., Ukai, S. and Kawashima, S., *Sur solutions à support compact de L'équation d' Euler compressible*, Japan J. Appl. Math., **3** (1986), 249–257.

[33] Alinhac, S., *Explosion géométrique pour des systémes quasi-linéaries*, Amer. J. Math., **117** (1995).

[34] Rauch, J., *BV estimates fail for most quasilinear hyperbolic systems in dimension greater than one*, Comm. Math. Phys. **106** (1986), 481–484.

[35] Ben-Dor, G., *Shock Wave Reflection Phenomena.* Springer-Verlag, New York, 1992.

[36] Matzumura, A. and Nishida, T., *The initial value problem for the equations of motion of viscous and heat-conductive gases*, J. Math. Kyoto Univ. **20** (1980), 67–104.

[37] Hoff, D., *Global solution of the Navier-Stokes equations for multi-dimensional compressible flow with discontinuous, initial data*, J. Diff. Eqns., **120** (1995), 215–254.

[38] Lions, P. L., *Existence globale de solutions pour les équations de Navier-Stokes compressible isentropiques*, C. R. Acad. Sci. Paris **316** (1993), 1333–1340.

[39] Xin, Z., *Blow-up of smooth solutions to the compressible Navier-Stokes equations with compact density*, to appear in Comm. Pure Appl. Math., 1997.

[40] Goodman, J., *Stability of viscous scalar shock fronts in several dimensions*, Trans. Amer. Math. Soc. **311** (1989), 683–695.

[41] Xin, Z., *Asymptotic stability of planar rarefaction waves for viscous conservation laws in several dimensions*, Trans. Amer. Math. Soc. **319** (1990), 805–820.

[42] Szepessy, A., *High frequency asymptotics for 2D viscous shocks*, preprint, 1997.

[43] Schlichting, H., *Boundary-Layer Theory*, 7th Edition, McGraw-Hill, 1987.

[44] Xin, Z. and Yanagisawa, T., *Zero viscosity limit of the linearized Navier-Stokes Equations for a compressible viscous fluid in the half plane*, to appear in Comm. Pure Appl. Math., 1997.

[45] Fife, P., *Considerations regarding the mathematical basis for Prandtl's boundary layer theory*, Arch. Rat. Mech. Anal., **38** (1967), 184–216.

[46] Oleinik, O., *The Prandtl system of equations in boundary layer theory*, Dokl. Akad. Nauk S.S.SR., **150** 4(3) (1963) 583–586.

[47] E, W., and Engquist, B., *Blow-up of solutions of the unsteady Prandtl's equations*, to appear in Comm. Pure Appl. Math., 1997.

[48] Sammartino, M. and Caflisch, R. E., *Zero viscosity limit for analytic solutions of the Navier-Stokes equations on a half-space*, I & II, preprints, 1996.

[49] Klainerman, S. and Majda, A., *Compressible and incompressible fluids*, Comm. Pure Appl. Math. **35** (1982), 632–653.

[50] Constantin, P. and Foias, C., *Navier-Stokes Equations*. Lecture in Mathematics, Chicago, The University of Chicago Press, 1988.

[51] Ladyzhenskaya, O. A., *The Mathematical Theory of Viscous Incompressible Flow*. New York, Gordon and Breach Science Publishers, 1969.

[52] Majda, A. and Bertozzi, M., *Vorticity and the Mathematical Theory of Incompressible Fluid Flow*, 1997.

[53] Caffarelli, L., Kohn, R. and Nirenberg, L., *Partial regularity of suitable weak solutions to the Navier-Stokes equations*, Comm. Pure Appl. Math., **35** (1982), 771–837.

[54] Wieger, M., *Decay and stability in Lp for strong solutions of Cauchy problems for the Navier-Stokes equations.* The Navier-Stokes equations, theory and numerical methods. Proceedings Oberwolfach (1988), eds. J. G. Heywood, et al., Lect. Notes Math. **1431**, Berlin, Heidelberg, New York; Springer 1990, 95–99.

[55] Ponce, G., Raike, R., Sideris, T. C. and Titi, E. S., *Global stability of large solutions to the 3D Navier-Stokes Equations*, Comm. Math. Phys. **159** (1994), 329–341.

[56] Diperna, R. and Majda, A., *Concentrations in regularizations for 2-D incompressible flow*, Comm. Pure Appl. Math., **60** (1987), 301–345.

[57] Diperna, R. and Majda, A., *Oscillations and concentrations in weak solutions of the incompressible fluid equations*, Comm. Math. Phys. **108**, (1987), 667–689.

[58] Ukhovskii, M. R. and Indovich, V. I., *Axially symmetric flows of ideal and viscous fluids filling the whole space*, J. Appl. Math. Mech., **32** (1968), 52–62.

[59] Moore, D. W., *The spontaneous appearance of a singularity in the shape of an evolving vortex sheet*, Proc. Roy. Soc. London Ser. A **365** (1979), 105–119.

[60] Delort, J. M., *Existence de nappes de tourbillon en dimension deux*, J. Amer. Math. Soc. **4** (1991), 553–586.

[61] Majda, A., *Remarks on weak solutions for vortex sheets with a distinguished sign*, Indiana Univ. Math. J. **42** (1993), 921–939.

[62] Liu, J.-G. and Xin, Z., *Convergence of vortex methods for weak solutions to the 2-D Euler equations with vortex sheets data*, Comm. Pure Appl. Math. **48** (1995), 611–628.

[63] Lin, F. H. and Lu. C., *Partial regularity of the dynamic system modeling the flow of liquid crystal*, Discrete Contin. Dynamic System, **2** (1996), 1–22.

[64] Tian, G. and Xin, Z., *Gradient estimation on Navier-Stokes Equation*, preprint, 1996.

[65] Tsai, T.-P., *On Leray's self-similar solutions of the Navier-Stokes equation satisfying local energy-estimates*, preprint, 1997.

[66] Grad, H., *Singular and nonuniform limits of the Boltzmann equations*, SIAM Amer. Math. Soc. Proceedings I, Transport Theory, 1969, 125–156.

[67] Caflisch, R. E., *The fluid dynamic limit of the nonlinear Boltzmann equation*, Comm. Pure Appl. Math. **32** (1979) 589–616.

[68] Kawashima, S., Matsumura, A. and Nishida, T., *On the fluid approximation to the Boltzmann equation at the level of the Navier-Stokes equation*, Comm. Math. Phys. **70** (1979), 97–124.

[69] Nishida, T., *Fluid dynamic limit of the nonlinear Boltzmann equation in the level of the compressible Euler equation*, Comm. Math. Phys. **61** (1978), 119–148.

[70] Bardos, C., Golse F. and Levermore, C. D., *Fluid dynamic limits of kinetic equations II: Convergence proofs for the Boltzmann equations*, Comm. Pure Appl., to appear.

[71] Xin, Z., *The fluid-dynamic limit of the Broadwell model of the nonlinear Boltzmann equations in the presence of shocks*, Comm. Pure Appl. Math., **44** (1991), 679–713.

[72] Liu, J.-G. and Xin. Z., *Boundary-layer behavior in the fluid-dynamic limit for a nonlinear model Boltzmann equation*, Arch. Rat. Mech. Anal., **135** (1996), 61–105.

[73] Liu, J.-G. and Xin, Z., *Kinetic and viscous boundary layers for the Broadwell equations*, Transport Th. Stat. Phys.

[74] Bardos, C., Caflisch, R. and Nicolaenko, B., *The Milne and Kramers problems for the Boltzmann equations of a hard sphere gas*, Comm. Pure Appl. Math., **39** (1986), 323-352.

[75] Wang, W.-C. and Xin, Z., *On small mean path limit of Broadwell model with discontinuous initial data, the centered rarefaction wave case*, preprint, 1997.

[76] Diperna, R. and Lions, P., *On the Cauchy problem for Boltzmann equations: Global existence and weak stability*, Ann. Math., **130** (1989), 321–366.

Geometry and Topology Books
from International Press
www.intlpress.com
PO Box 38-2872, Cambridge, MA 02238-2872

Chern, A Great Geometer, *edited by S.-T. Yau* New expanded edition!!

Algebraic Geometry and Related Topics, *edited by J.-H. Yang, Y. Namikawa and K. Ueno.*

Differential Geometry, *by R. Schoen and S.-T. Yau*

Elliptic Curves, Modular Forms and Fermat's Last Theorem, NEW EXPANDED EDITION!! *edited by J. Coates and S.-T. Yau.*

Geometric Analysis and the Calculus of Variations, for Stefan Hildebrandt, *edited by Jürgen Jost*

Geometric Topology, Proceedings of the Georgia

Geometry and Topology Conference, *edited by W. Kazez*

Geometry, Topology, and Physics for Raoul Bott, *edited by S.-T. Yau*
Goukova Geometry and Topology Conferences, 1993, 1994, 1995, 1996, *edited by S. Akbulut, T. Onder and R. Stern*

Geometric Analysis and the Calculus of Variations, for Stefan Hildebrandt *edited by Jürgen Jost*

Lectures on Harmonic Maps, *by R. Schoen and S.-T. Yau*

Lectures on Low Dimensional Topology, *edited by K. Johannson*

Lectures on Nonlinear Wave Equations, *by Christopher D. Sogge*

L^2 Moduli Spaces on 4-Manifolds with Cylindrical Ends, *by C. Taubes*

The L^2 Moduli Space and a Vanishing Theorem for Donaldson Polynomial Invariants, *by J. Morgan, T. Mrowka, D. Ruberman*

Mathematical Physics Books
from International Press
www.intlpress.com
PO Box 38-2872, Cambridge, MA 02238-2872

Quantum Groups: from Coalgebras to Drinfeld Algebras *by
Steven Shnider and Shlomo Sternberg*

75 years of Radon Transform *by Simon Gindikin and Peter
Melchior*

Perspectives in Mathematics Physics *by Robert Penner and S.T.
Yau*

**Proceedings of the International Conference on Mathematical
Physics** XIth volume edited by Daniel Iagolnitzer, XIIth volume
edited by David DeWitt

Mirror Symmetry volume 1 edited by S.T.Yau, volume 2 edited
by Brian Greene

Physics Series
Physics of the Electron Solid *by S.-T. Chui*

**Proceedings of the Second International Conference on
Computational Physics,** *by D.H. Fenb and T.-Y Zhang*

Chen Ning Yang, a Great Physicist of the Twentieth Century,
by S.-T. Yau

Yukawa Couplings and the Origins of Mass, *edited by P.
Ramond*

Textbooks from International Press

www.intlpress.com

PO Box 38-2872, Cambridge, MA 02238-2872

Calculus: A Computer Algebra Approach,
by L. Ashel and D. Goldfeld

Uniquely designed for use with computer algebra systems and sophisticated calculators, this course also works well with a computer laboratory. The students are encouraged to use technology for manual computation while they rapidly progress through the concepts of differential and integral calculus, mathematical modeling and optimization, ordinary differential equations, differential calculus for vector valued and multi-variable functions. The students will progress to vector geometry and coordinate systems, two and three dimensional graphical display, multiple integration, vector fields and line integrals, and on to Fourier series and the Fourier expansion theorem.

Basic Partial Differential Equations,
by D. Bleecker and G. Csordas

Using partial differential equations as a tool to predict systems based on underlying physical principles, this text is accessible to students with calculus background without presupposing courses in linear algebra or ordinary differential equations. The text takes an inductive approach, motivating students to see the relevance and applications of the general principles which govern the subject, but rigorous proofs are still available.

The problems included range from routine to very challenging. The problems are constructed to require genuine understanding of principles as well as procedures, and the solutions manual is available only to instructors—not to the general public. Suggestions from the author indicate how the book can be used in one semester, two quarter, or two semester courses.

A First Course in Differential Geometry,
by C.-C. Hsiung

This text treats the traditional topics of curves and surfaces in 3-dimensional Euclidean space with attention equally on local and global properties. Extension to higher dimensional spaces and more general surfaces is available. In addition to analytical approaches, the text gives geometric interpretation to its subjects and encourages the development of geometric intuition. The book uses vector analysis and exterior differential calculus but does not require tensor calculus.

Part 1 reviews point-set topology, advanced calculus, and linear algebra. Part 2 establishes a general local theory for curves in three dimensional Euclidean space including proving the uniqueness theorem for curves in E^3. Part 3 develops a Local theory for surfaces in E^3, and Part 4 covers orientation of surfaces, surfaces of constant Gaussian curvature, and global theorems for surfaces.

Proceedings of the International Seminar on Singularities and Complex Geometry, *edited by Qi-Keng Lu, Stephen Yau and Anatoly Libgober*

Surveys in Differential Geometry, Volume I, Volume II, *edited by C.C. Hsiung and S.-T. Yau,* Volume III, *edited by Chuu Lian Terng and Karen Uhlenbeck.*

Topics in Symplectic 4-manifolds, volume 1, First International Press Lecture Series, *edited by Ronald Stern*

Tsing Hua Lectures on Geometry and Analysis, *edited by S.-T. Yau*

Current Developments in Mathematics
Proceedings of the annual conference sponsored by Harvard and MIT

1995 Proceedings: lectures by Henri Darmon, Fred Diamond, and Richard Taylor, Mikhail Lyubich, Ib Madsen, Curtis McMullen, and Gang Tian

1996 Proceedings: lectures by Richard Borcherds, Gerrit Heckman and Eric Opdam, Ehud Hrushovski,k and Yves Meyer

1997 Proceedings: lectures by A. Connes, L. C. Evans, P. Sarnak, and W. Soergel and open problems by A. Jaffe, D. Stroock, B. Mazur, C. Taubes, and D. Jerison